Managing Temperature Effects in Nanoscale Adaptive Systems

T0137367

David Wolpert • Paul Ampadu

Managing Temperature Effects in Nanoscale Adaptive Systems

 Springer

David Wolpert
University of Rochester
Rochester, NY, USA
david.s.wolpert@gmail.com

Paul Ampadu
University of Rochester
Rochester, NY, USA
ampadu@ece.rochester.edu

ISBN 978-1-4899-8722-8   ISBN 978-1-4614-0748-5 (eBook)
DOI 10.1007/978-1-4614-0748-5
Springer New York Dordrecht Heidelberg London

Springer is part of Springer Science+Business Media (www.springer.com)

"To Joshua, Julie, Gloria, and Brian"
David Wolpert

"To Luzviminda, for her courage;
To Luzann, for her love;
To Majelia and Paul Jr, for their faith."
Paul Ampadu

Preface

The primary objective of this book is to highlight the growing importance of temperature-aware circuit and system design. While previous books have discussed the impact of temperature on nanoscale systems, we provide a holistic discussion of temperature management including physical phenomena (reversal of the MOSFET temperature dependence) that are only recently becoming problematic, including circuit techniques for detecting, controlling, and adapting to these phenomena. To complete the discussion we also provide details on the general aspects of thermal-aware system design and management of temperature-induced faults. This book is based on the research carried out by David Wolpert from 2005 to 2011 in the EdISon Research Group at the University of Rochester during doctoral studies under the supervision of Professor Paul Ampadu.

Temperature variations affect system speed, power, and reliability by altering device parameters such as threshold voltage (V_T), mobility (μ), and saturation velocity (v_{sat}). The impact of temperature on device performance changes as technology scales. Device on-current has generally been known to *decrease* as temperature increases; however, as technologies scale further into the nanometer regime, the changes in device parameters and their temperature dependences can cause on-current to *increase* as temperature increases under certain conditions. In addition to device current changing with temperature, careful control of threshold and supply voltages can render device on-current nearly insensitive to changes in temperature. This dissertation examines the mechanisms affecting the temperature dependence of device current in nanoscale systems, and proposes a set of techniques for (i) detecting the temperature dependence, (ii) controlling and exploiting the temperature dependence, and (iii) compensating for temperature-induced reliability issues.

Detection of the temperature dependence will become increasingly critical as technology scales and the impact of temperature on device current reverses at near-nominal voltages. Existing temperature sensors are designed assuming that device current decreases as temperature increases; thus, the reversal of the temperature dependence will cause problems such as false positives, undetected overheating, or undetected timing failures. In this dissertation, we propose a new type of sensor system that can determine the temperature dependence as well as the operating temperature; this sensor system ensures correct detection of overheating and timing-related errors regardless of the temperature dependence, improving system reliability.

To control the temperature dependence, prior work has examined the use of multi-V_T design methodologies, adaptive body bias (ABB) methods to control V_T, and supply voltage scaling to a technology-specific temperature-insensitive supply voltage (V_{INS}). Unfortunately, the use of V_{INS}—even with multi-V_T devices and ABB—restricts design to a very specific delay and power operating point, preventing the use of common adaptive techniques such as dynamic voltage scaling. Furthermore, NMOS and PMOS devices each have separate values of V_{INS}, limiting the effectiveness of 'temperature-insensitive' design. In this dissertation, we propose a new method of controlling a circuit's temperature dependence using programmable temperature compensation devices to individually tune pull-up and pull-down networks to their temperature-insensitive operating points. The proposed method also extends the range of supply voltages that can be made temperature-insensitive, achieving insensitivity at nominal voltage for the first time.

Although temperature dependences are generally considered to be undesirable, in some applications these dependences can actually be exploited to improve performance. For example, long interconnect links are commonly operated at reduced supply voltages to save energy, while the transmitter and receiver units operate at higher voltages. We propose a delay-borrowing method to exploit the different temperature dependences in the link and transceiver, dramatically improving both energy performance and link reliability.

Despite the immense efforts of circuit designers to guardband their systems and maintain reliability in the presence of temperature variations, temperature issues still result in transient effects like temperature-induced delay uncertainty and timing failures, as well as permanent faults caused by hot-electron effects or increased electromigration. In this dissertation, we present methods of managing these reliability issues in a variety of applications, such as improving delay uncertainty in clock trees, integrating temperature-awareness into an adaptive multi-core control unit, and using an in-line test system to bypass intermittent and permanent errors in on-chip interconnect links.

Rochester, NY   David Wolpert
                                                                                Paul Ampadu

Acknowledgments

The authors would like to thank Chuck Glaser from Springer for his encouragement and assistance. Thanks also to our outstanding EdISon Research Group alumni Dr. Bo Fu and Dr. Qiaoyan Yu for their support, criticisms, and friendship throughout the years this research was being performed. We would also like to thank Dr. Teijo Lehtonen from the University of Turku in Finland for his extensive contributions to the work presented in Chap. 7.

The research presented in this book was made possible in part by the U.S. National Science Foundation under Grant ECCS-0925993 and CAREER Award ECCS-0954999, as well as grants from the New York State Foundation for Science, Technology and Innovation, the Air Force Office of Scientific Research, the Office of Naval Research, and the Semiconductor Research Corporation.

Contents

1 **The Role of Temperature in Electronic Design** 1
  1.1 Temperature and Reliability in Nanoscale System Design 4
  1.2 Global and Local Temperature Variation 7
  1.3 Thermal Control in VLSI Systems 8
  1.4 Book Overview 10
  References 11

2 **Temperature Effects in Semiconductors** 15
  2.1 Material Temperature Dependences 15
    2.1.1 Energy Band Gap 15
    2.1.2 Carrier Density 16
    2.1.3 Mobility 18
    2.1.4 Carrier Diffusion 20
    2.1.5 Velocity Saturation 20
    2.1.6 Current Density 21
    2.1.7 Threshold Voltage 22
    2.1.8 Leakage Current 22
    2.1.9 Interconnect Resistance 24
    2.1.10 Electromigration 24
  2.2 Normal and Reverse Temperature Dependence 25
    2.2.1 Discovery of the Normal and Reverse
        Temperature Dependences 26
  2.3 Temperature and Technology Scaling 28
    2.3.1 The Reverse Temperature Effect
        and High-κ/Metal Gate Technology 28
  References 32

3 **Sensing Temperature Dependence** 35
  3.1 Low Overhead, Energy-Efficient Adaptive Temperature Sensor .. 35
    3.1.1 Temperature Sensor Design 36
    3.1.2 Sensor Characterization and Results 40

3.1.3   Adapting Sampling Rate and Resolution 43
3.1.4   Process Compensation Unit 45
3.2   A Temperature Dependence Sensor 47
3.2.1   Temperature Dependence Sensor Design 47
3.2.2   Component Temperature Sensors 50
3.2.3   Dependence Sensor Characterization 51
3.3   Fabricated Temperature Dependence Sensor Chip 54
3.3.1   Measurement of Sensor System Functionality 55
3.3.2   Measurement of Sensor System Accuracy
        and Process Compensation 56
3.3.3   Comparison with Other Temperature Sensors 59
3.4   Summary 59
References 60

4   Variation-Tolerant Adaptive Voltage Systems 63
4.1   Reliability Issues in Nanoscale Systems 64
4.1.1   Process Variation 65
4.1.2   Runtime Variation 66
4.1.3   Temperature Variation Impact 67
4.1.4   Voltage Variation Impact 68
4.1.5   Combined Effects of Variation 69
4.2   Design Considerations 70
4.2.1   Percentage-Based Design for Variation 71
4.2.2   Yield-Based Design for Variation 72
4.2.3   Optimizing Performance with Multiple Constraints 74
4.2.4   Temperature-Robust Performance Yield
        through Supply Voltage Selection 74
4.3   Adaptive Delay Correction in Adaptive
      Voltage Scaling Systems 76
4.3.1   Defining the Operating Voltages 77
4.3.2   Variation-Tolerant Frequency Guardbands 78
4.3.3   Proposed Adaptive System Design 79
4.3.4   Results and Comparison with Prior Work 80
4.4   Multi-Core System Design with Variation Tolerance
      and a Low Power Safety Mode 85
4.4.1   Multi-Core Framework 85
4.4.2   Multi-Core Simulation Results 87
References 90

5   Controlling the Temperature Dependence 93
5.1   Existing Methods for Reducing Temperature Sensitivity 94
5.1.1   Temperature Insensitivity with Multi-Threshold Design .. 95
5.1.2   Temperature Insensitivity with Adaptive
        Body Biasing (ABB) 96
5.2   Proposed Approach for Temperature Insensitivity 97

		5.2.1	PTCD Inverter	101
		5.2.2	Comparison of Temperature-Insensitive Voltage Ranges	104
		5.2.3	PTCD Integration with Other Logic Structures	105
	5.3	Applications		109
		5.3.1	Temperature-Insensitive Clock Tree	109
		5.3.2	Temperature Sensitivity Adjustment for Improved Sensor Accuracy	110
	5.4	Discussion		113
		5.4.1	Layout Area Overhead	113
		5.4.2	Impact of Sizing on Temperature Insensitivity	114
		5.4.3	Impact of Slew Rate on Temperature Sensitivity	115
		5.4.4	Combined Conventional CMOS/PTCD Datapaths	116
		5.4.5	Impact of Wire Temperature Dependence on PTCD Methodology	116
		5.4.6	Variation Considerations	117
	5.5	Compensation for Process Variations and Aging		119
	5.6	Summary		120
	References			122
6	**Exploiting Temperature Dependence in Low-Swing Interconnect Links**			**125**
	6.1	Related Work		126
	6.2	Temperature-Aware Low Voltage Link Design		127
	6.3	Results		131
		6.3.1	Simulation Setup	131
		6.3.2	System Characterization	132
		6.3.3	Comparison with Conventional Low-Swing Link	133
		6.3.4	Comparison with Temperature-Insensitive Operation	134
		6.3.5	Area Overhead	136
	6.4	Discussion		137
		6.4.1	Integration of Multiple Temperature Sensors	137
		6.4.2	Integration with Error Control Coding	137
		6.4.3	Level Shifter Design for Systems with Multiple Temperature Dependences	138
	6.5	Summary		139
	References			139
7	**Avoiding Temperature-Induced Errors in On-Chip Interconnects**			**141**
	7.1	The Need for Error Protection in On-Chip Interconnects		142
		7.1.1	Transient Errors	142
		7.1.2	Intermittent and Permanent Errors	142
	7.2	Error Control Coding Fundamentals		143
		7.2.1	Triple Modular Redundancy	143
		7.2.2	Hamming Codes	144

       7.2.3   Interleaving...   145

  7.3   Related Work   145

  7.4   Permanent Error Correction   146

  7.5   In-Line Test (ILT) Error Detection Method......................   147

       7.5.1   In-Line Test Procedure....................................   147

       7.5.2   Analysis of Open and Shorted Wires......................   148

  7.6   Adaptive Correction Framework   152

       7.6.1   Encoder and Decoder   153

       7.6.2   Reconfiguration Units.....................................   153

       7.6.3   Reconfiguration Unit Control.............................   154

       7.6.4   Transmission Protocol   155

  7.7   Case Study and Comparison with Prior Work....................   158

       7.7.1   Implementation Results...................................   158

       7.7.2   Error Tolerance Analysis   161

  7.8   Discussion...   164

  References..   166

8   Future Work and Open Problems   169

Index...   171

List of Symbols

n	Electron concentration gradient
p	Hole concentration gradient
α	Technology-specific exponent for drain current
α_E	Material-specific constant for Varshni equation (eV/K)
α_R	Temperature coefficient of resistance
α_{sw}	Switching activity factor
α_{th}	Thermal diffusivity (cm^2/s)
α_{vsat}	Saturation velocity temperature coefficient
α_{VT}	Threshold voltage temperature coefficient
α_μ	Mobility temperature exponent
β	Sizing ratio of a MOSFET gate (ratio of PMOS size to NMOS size)
β_E	Material-specific constant for Varshni equation (K)
γ	Body effect parameter
ΔT	Change in temperature
ΔV	Change in voltage
ε	Permittivity (F/m)
ε_{Si}	Relative permittivity of Silicon (11.7)
η	Device constant coefficient for inversion layer change density
κ	Dielectric constant
μ	Mobility (cm^2/V·s)
μ_0	Mobility at nominal temperature
μ_{cb}	Bulk charge Coulombic scattering component of mobility
μ_{eff}	Effective mobility
μ_{int}	Interface charge Coulombic scattering component of mobility
μ_m	Arithmetic mean
μ_n	Electron mobility
μ_p	Hole mobility
μ_{ph}	Phonon scattering component of mobility

μ_{sr}	Surface roughness scattering component of mobility
ξ	Electric field (V/m)
ξ_{eff}	Effective transverse electric field
ρ	Substrate density (g/cm^3)
σ	Standard deviation
τ	Delay
ϕ_F	Fermi energy
Φ_{GC}	Gate-channel work function difference
ϕ_{gs}	Gate-substrate contact potential
ϕ_T	Thermal voltage ($= kT/q$)
A	Constant in Shockley diode model
A_j	Constant in Black's equation
Al	Aluminum
C	Capacitance
c	Specific heat (J/g°C)
C_c	Coupling capacitance
C_L	Load capacitance
Clk→Q delay	Delay from clock signal edge to output of flip-flop
Clk	Clock
C_{ox}	Oxide capacitance
C_s	Substrate capacitance
Cu	Copper
D	Diffusion current
d_{min}	Minimum Hamming distance between codewords
D_n	Electron diffusion current
D_p	Hole diffusion current
E	Energy
E_a	Activation energy
E_C	Conduction band energy level
E_F	Fermi energy level
E_g	Energy band gap
E_{g0}	Energy band gap at T=0 K
En	Sensor enable period
E_V	Valence band energy level
f	Frequency
f_{base}	Baseline frequency
Fe	Iron
FF	Fast PMOS/Fast NMOS process corner
$f_{guardband}$	Guardbanded frequency
$f_{osc,max}$	Maximum oscillator frequency over a given temperature range
$f_{osc,min}$	Minimum oscillator frequency over a given temperature range
f_{osc}	Oscillator frequency
GaAs	Gallium Arsenide
Ge	Germanium

Ham	Hamming codes
I_0	Reverse saturation current in Shockley diode model
$I_{125°C}$	Current at 125°C
$I_{-55°C}$	Current at −55°C
I_D	Drain current
I_{leak}	Leakage current
I_{peak}	Peak current during short-circuit period
IR Drop	Drop in voltage along an interconnect associated with finite wire resistance
I_{sub}	Subthreshold leakage current
I-T dependence	Dependence of device current on temperature
J	Current density (A/m^2)
J_N	Electron current density
J_P	Hole current density
k	Boltzmann constant ($1.38 \cdot 10^{-23}$ J/K)
k_{th}	Thermal conductivity (W/cm°C)
L	Device length
L_x	Length of unit under test
n	Electron density
n_j	Constant scaling factor in Black's equation
N_A	Acceptor dopant concentration
N_C	Density of states in the conduction band
N_D	Donor dopant concentration
N_{FF}	Number of flip-flops
N_G	Gate dopant concentration
n_i	Intrinsic carrier concentration
Ni	Nickel
$N_{pulse,max}$	Maximum number of pulses generated by a pulse generator over a given temperature range
$N_{pulse,min}$	Minimum number of pulses generated by a pulse generator over a given temperature range
N_{pulse}	Number of pulses generated by a pulse generator
N_V	Density of states in the valence band
p	Hole density
P	Power dissipation
P_{avg}	Average power dissipation
P_{dyn}	Dynamic power dissipation
P_{err}	Probability of correct transmission in the presence of err errors
P_{idle}	Power dissipation when unit is idle
P_{leak}	Leakage power dissipation
P_s	Technology-specific constant for velocity-saturated drain current
P_{sample}	Power dissipation per sample
P_{sc}	Short-circuit power dissipation

$PTCD_{nset}$	Programmable Temperature Compensation Device technique using additional devices only in the pull-down path
$PTCD_{nset+pset}$	Programmable Temperature Compensation Device technique using additional devices in both the pull-up and pull-down paths
PW	Enable pulse width
P_{perm}^{tran}	Probability of correct transmission in the presence of *perm* permanent and *tran* transient errors
q	Charge on an electron ($1.6 \cdot 10^{-19}$ C)
Q_b	Substrate depletion charge density
Q_{B0}	Depletion region charge density at surface inversion
Q_{crit}	Critical charge
Q_{inv}	Inversion layer charge density
Q_{ox}	Oxide-substrate interface charge density
Q_{ss}	Surface charge density
q_v	Power density per unit volume (W/cm^3)
r	Distance between heat source and measurement point in thermal system
R	Resistance
R_0	Resistance at nominal temperature
R_{avg}	Average sensor resolution over a given temperature range
RC model	Interconnect model including resistive and capacitive components
R_{min}	Minimum sensor resolution over a given temperature range
R_{open}	Open resistance between two wires
r_{sample}	Sensor sampling rate
R_{short}	Short resistance between two wires
R_{th}	Thermal resistance (°C/W)
s	Number of spare wires in an interconnect link
S	Technology scaling factor
Si	Silicon
SiC	Silicon carbide
SiO_2	Silicon dioxide
SS	Slow PMOS/Slow NMOS process corner
T	Temperature
t	Time
t'	Time step
T_0	Nominal temperature (baseline temperature used for empirical modeling)
T_a	Ambient temperature
$T_{accuracy}$	Accuracy of a temperature sensor
T_{chip}	Actual chip temperature
T_{clk}	Clock period
$t_{correction}$	Adaptation latency of a temperature-compensation system
$t_{d,125°C}$	Delay at 125°C
$t_{d,-55°C}$	Delay at −55°C

$t_{d,f}$	Delay of a falling edge transition
$t_{d,r}$	Delay of a rising edge transition
t_{enable}	Enable pulse width of a temperature sensor
$T_{guardband}$	Temperature guardband
t_h	Hold delay
T_j	Junction temperature
t_L	Link delay
$T_{measured}$	Measured chip temperature
T_{OP}	Temperature reading taken at V_{OP}
T_{ox}	MOSFET gate oxide thickness
t_{R_RX}	Shortest delay through ILT reconfiguration logic in the receiver
t_{R_TX}	Shortest delay through ILT reconfiguration logic in the transmitter
$T_{range,max}$	Maximum temperature of the desired temperature range
$T_{range,min}$	Minimum temperature of the desired temperature range
T_{REF}	Temperature reading taken at V_{REF}
$t_{response}$	Response time of a temperature sensor
t_{sc}	Short-circuit current period
t_{sensor}	Sensor polling time
TT	Typical PMOS/Typical NMOS process corner
$T_{threshold}$	Threshold temperature used to trigger an adaptive system
U	Ratio of supply voltages for scaling purposes
$V_{B,N}$	NMOS body voltage
$V_{B,P}$	PMOS body voltage
V_b	MOSFET body voltage
v_d	Drift velocity
V_d	MOSFET drain voltage
V_{DD}	Supply voltage
V_{ds}	MOSFET drain-source voltage $(= V_d - V_s)$
V_{FB}	Flat band voltage
V_g	MOSFET gate voltage
V_{gs}	MOSFET gate-source voltage $(= V_g - V_s)$
V_{High}	Higher of two supply voltages
$V_{INS,N}$	Temperature insensitive voltage in NMOS devices
$V_{INS,P}$	Temperature insensitive voltage in PMOS devices
V_{INS}	Temperature-insensitive voltage
V_{Low}	Lower of two supply voltages
V_{NOM}	Technology-specified nominal voltage
V_{OP}	Operating voltage of a unit being monitored
V_{REF}	Reference voltage
V_s	MOSFET source voltage
v_{sat}	Saturation velocity
v_{sat0}	Saturation velocity at nominal temperature
V_{sb}	MOSFET source-body voltage $(= V_s - V_b)$
V_{swing}	Swing voltage

$V_{T,VB0}$	Unbiased threshold voltage
V_T	Threshold voltage
V_{T0}	Threshold voltage at nominal temperature
W	Device width
W_x	Width of unit under test
x	Cartesian dimension
y	Cartesian dimension
z	Cartesian dimension

List of Acronyms

+I-T Slope	Reverse temperature dependence region
ABB	Adaptive body bias
ADC	Analog-to-digital converter
AOI	AND-OR-INVERT
AVS	Adaptive Voltage Scaling
BCH	Bose-Chaudhuri-Hocquenghem codes
BSIM	Berkeley Short-channel IGFET Model
DVFS	Dynamic Voltage and Frequency Scaling
DVS	Dynamic Voltage Scaling
DVS+DB	Dynamic Voltage Scaling with Delay Borrowing
ECC	Error Control Coding
EUV Lithography	Extreme ultraviolet lithography
FBB	Forward body bias
FO4	Fan-out of 4, used to signify a consistent load based on technology parameters
GALS	Globally-Asynchronous, Locally Synchronous
HVT	High threshold voltage
I/O	Input/Output
IC	Integrated Circuit
IGFET	Insulated-Gate Field Effect Transistor
ILT	In-Line Test system for interconnect error detection
–I-T Slope	Normal temperature dependence region
LAGS	Locally-Asynchronous, Globally Synchronous
LCFD	Level Converter with Frequency Doubler
LUT	Look-Up Table
LVDS	Low-Voltage Differential Signaling
LVT	Low threshold voltage
MBU	Multi-Bit Upset
MOSFET	Metal-Oxide-Semiconductor Field-Effect Transistor
MTTF	Mean time to failure

ND	Normal temperature dependence region
NMOS	N-type Metal-Oxide-Semiconductor Field-Effect Transistor
NoC	Network-on-Chip
OAI	OR-AND-INVERT
PDN	Pull-down network
PLL	Phase-Locked Loop
PMOS	P-type Metal-Oxide-Semiconductor Field-Effect Transistor
PTCD	Programmable Temperature Compensation Device
PUN	Pull-up network
PV var	Process and voltage variation
PVT	Process, Voltage, and Temperature
RD	Reverse temperature dependence region
ROM	Read-Only Memory
SBB	Standard body bias
SEU	Singe-Event Upset
SoI	Silicon-on-Insulator
SSD	Syndrome Storing-based Detection system for interconnect error detection
SVT	Standard threshold voltage
TMR	Triple Modular Redundancy
TPG	Test Pattern Generator
VLSI	Very Large Scale Integration
W.C	Overall worst case behavior
W.C./V	Worst case behavior specified by a voltage point
$\Sigma\Delta$ ADC	"Sum of parts" analog-to-digital conversion

Chapter 1
The Role of Temperature in Electronic Design

Four hundred years ago (c. 1600 CE), a bearded old man added a new contraption to his workshop—a hollow glass bulb attached to a long, hollow glass tube. He warmed the bulb in his hands and lowered the open end of the tube into a cool liquid. As the air inside the bulb cooled, some liquid was drawn upward into the instrument. The warmer the man could make the bulb before placing the tube in the liquid, the further up the tube the liquid would climb. The man's name was Galileo Galilei, and he was experimenting with a new invention: the thermoscope [1].

While Galileo is believed to be the first inventor of the thermoscope, only second-hand accounts of his work have survived. The first known published image of a thermoscope was in 1620 by Santorio Sanctorius [2], shown in Fig. 1.1a, although a similar contraption was created by two other inventors in the same time period. One of these inventors, Robert Fludd, expanded upon an observation made by the ancient Greek Philo of Byzantium (~200 BCE) who observed that when a glass jug was placed upside-down in water and heated by the sun, bubbles of air would emerge from under the lip. Fludd's contribution was to place markings on the tube to indicate different temperatures, creating the first thermometer. The other inventor, Cornelius Drebbel (better known for the two-lens microscope and the submarine), created a similar contraption in the same time period, shown in Fig. 1.1b [3].

A 100 years later (c. 1720 CE), Daniel Gabriel Fahrenheit made mercury and alcohol thermometers popular, producing some of the first standard, calibrated thermometers. Before Fahrenheit's time, there was no established scale by which a temperature could be measured. Because there was no common scale, comparing a temperature between two locations required two thermometers crafted and calibrated by the same instrument maker. Thermometers made by different people would provide different readings, with the lowest point often marked on the coldest day of the year and the highest point marked on the warmest. Ole Rømer (better known for discovering light's finite speed) was the first to use the freezing and boiling points of water as frames of reference in 1702 CE. Fahrenheit standardized Rømer's system to create a more easily labeled scale, with 64° separating the

D. Wolpert and P. Ampadu, *Managing Temperature Effects in Nanoscale Adaptive Systems*, DOI 10.1007/978-1-4614-0748-5_1,
© Springer Science+Business Media, LLC 2012

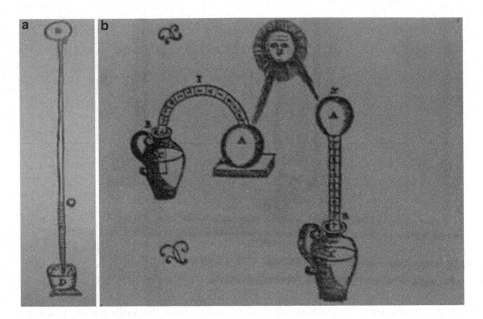

Fig. 1.1 (a) Sanctorius' thermoscope [2], (b) Fludd's thermometer [3]

freezing point of water and human body temperature, and 32° separating the freezing point of a brine solution and the freezing point of water (64 equidistant notches between 32°F and 96°F can be easily reproduced by halving the endpoints six—five times for 0–32°F) [2]. Fahrenheit's final scale (used today) adjusted the value of the degree to set the difference between freezing water and boiling water to 180°F; the resulting change in the value of the degree increased body temperature on the Fahrenheit scale to the present value of 98.6°F.

Anders Celsius refined Fahrenheit's work later in the 1700s to use a simpler number system, with 100° separating the freezing and boiling points of water. Celsius confirmed these points to be independent of latitude and dependent on atmospheric pressure. In 1848 CE, William Thomson (Lord Kelvin) proposed the need for a scale on which zero degrees corresponded to "infinite cold" (referred to today as "absolute zero"). Kelvin used Celsius degrees as an increment, with room temperature set to 300 K. The Kelvin scale is today the most widely used by scientists; temperatures as low as 250 pK ($2.5*10^{-10}$ K) have been reported [4].

The measurement of temperature has come a long way from its humble beginnings. Today, temperature sensors are prevalent in all aspects of society, from meteorology to medicine to microprocessors. The ability to measure temperature (using mercury thermometers) comes from our understanding of how temperature affects the expansion and contraction of mercury. To ensure that our electronics are reliable despite changes in temperature, we must examine how temperature affects those materials and what impact those effects have on our system designs.

Fig. 1.2 Archimedes' burning glass warfare [7]

Temperature has been known to cause changes in materials as far back as the early Stone Age some 800,000 years ago (the earliest surviving evidence of a Homo Erectus site with concentrations of charred flint items) [5]. Among the most famous historical uses of temperature effects was by Archimedes (~250 BCE), who used an array of mirrors to ignite attacking ships [6] (shown by the beautiful fresco in Fig. 1.2 [7]). In 1600 CE, William Gilbert (who coined the latin term 'Electricus', which later became electricity) described how increasing temperature reduced the attraction between oppositely charged objects [8]. In 1821 CE, Sir Humphry Davy presented to the Royal Society of London the results of an experiment in which [9] "the conducting power of metallic bodies varied with the temperature, and was lower, in some inverse ratio, as the temperature was higher."

This has come to be known as the *normal* temperature dependence of a material, and is discussed in detail in Chap. 2.

One year later, in 1822 CE, Thomas Johann Seebeck noted that when the junction temperature of two dissimilar metals was increased, a current was created. This is now known as the Seebeck or thermoelectric effect [10]. A similar principle was discovered by Jean Charles Athanase Peltier in 1834 CE, who found that when a current flows through such a junction, heat is absorbed from one end of the junction and moved to the other [11]. This effect is material dependent; the material's majority carrier determines whether heat flows in the same direction as the current or in the opposite direction. The Peltier effect (or Peltier–Seebeck effect) is now commonly used in electronic heat-transfer and cooling systems, including small electric refrigerators. The Peltier and Seebeck effects have also been examined for regulating chip temperature and recycling thermal energy back into electrical energy [12].

One of the most important temperature-related material discoveries was in 1833, when Michael Faraday discovered that silver sulphide was conductive at high temperatures and nearly insulating at low temperatures [13]. This was in direct contrast to the temperature dependence observed in metals, which become less conductive as temperature is increased. Faraday had discovered the semiconductor, and the reasoning behind this new relationship between temperature and current was later linked to the thermal excitation of carriers, described in Chap. 2. Note that Faraday and Davy's results are not contradictory; Davy's result describes the conductance of a material given an abundance of carriers, and Faraday's result describes the change in conductance when carriers are absent or present from thermal excitation.

1.1   Temperature and Reliability in Nanoscale System Design

The findings of Davy, Seebeck, Peltier, and Faraday have laid the foundation for the impact of temperature on electronic systems. Although temperature is one of many sources of variation facing nanoscale systems [14], its tight coupling with power dissipation and power density makes it among the most important of factors constraining nanoscale system design. Changes in temperature can have catastrophic effects on system performance and functionality, and are becoming increasingly problematic as technologies scale. Power dissipation in a material is related to temperature by the material's thermal conductivity, measured in Watts per meter Kelvin (W/(m·K)). Power dissipation and associated temperature consequences have been described as the major limitation which will end device scaling many generations before fundamental atomic limits are reached [15, 16].

As shown by Davy and Faraday, changes in temperature affect the conductivity of the material, which affects the speed of computation in computing systems. When temperatures increase beyond some maximum tolerance, a chip can no longer function at its required speed (a notable exception to this behavior is described in Chap. 2), resulting in erroneous computations. As power dissipations increase beyond our ability to distribute and remove the generated heat, local temperatures will continue to increase, worsening the potential for delay failure (not to mention thermal failure, when the temperatures are sufficient to melt some compounds).

Power management systems are critical for reducing the aforementioned design issues. Indeed, the 2007 International Technology Roadmap for Semiconductors (ITRS) states [17] "In addition to the $2\times$ increase in transistor count per generation, power management is now the primary issue across most application segments." Although the power dissipated by each transistor decreases by $1/S$ [18] (where S is the technology scaling factor, ~0.7) with each new technology generation, the net power dissipation in each new microprocessor can actually increase as more transistors are added to increase functionality. These trends are shown in Fig. 1.3,

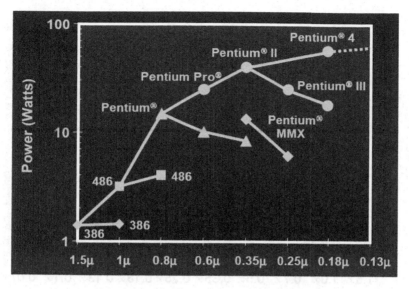

Fig. 1.3 Power dissipation across multiple generations of Intel chips [19]

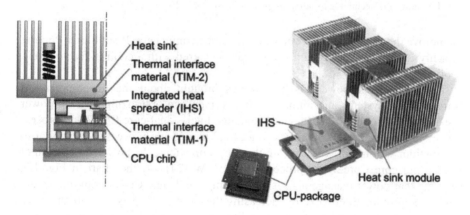

Fig. 1.4 CPU cooling system [22]

where each labeled point is a new chip generation and the branches shown are the changes in power dissipation as chips are scaled to smaller technologies.

The thermal conductivity of Si is fixed, thus the increase in power with each new chip generation requires more advanced cooling systems to limit the increases in chip temperatures. Thermal conductivity is even lower in silicon-on-insulator (SoI) technologies than in bulk Si technologies, making temperature more difficult to manage [20, 21]. A common implementation of a passive cooling system is shown in Fig. 1.4, including the chip, heat spreader, and heat sink. Thermal interface material is a high thermal conductivity substance used

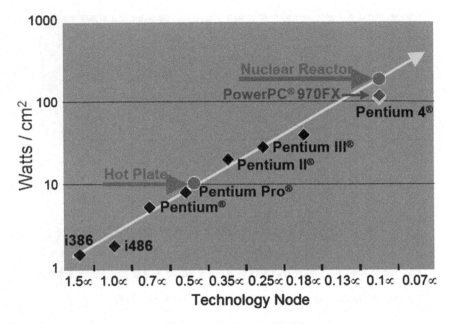

Fig. 1.5 Impact of technology scaling on power density [25, 26]

to improve the flow of heat between each separate component. If additional cooling capacity is required, the system shown can be paired with a cooling fan or liquid cooling unit.

The problem of power dissipation has brought about the development of better cooling systems and immense improvements in optimizing on-chip power dissipation. To highlight recent achievements, Intel's dual-core Core2Duo processor dissipates just 65 W under maximum load in a 65 nm process [23], similar to the Pentium 4 from Fig. 1.3. Despite these optimizations, the power dissipation of Intel's four-core Core2Quad processor is 136 W [24], off the chart in Fig. 1.3, showing that power management is an ongoing challenge. Considering the recent trend of improving throughput by increasing the number of cores on chip, there is a growing need for techniques to manage these power dissipations and their temperature effects.

Assuming that the electric field ξ is held constant with scaling, each transistor's power density (measured in W/cm^2) should not change between technology generations; unfortunately, the scaling trends for supply voltage and electric field are unable to keep up with the scaling of device sizes, and power densities have been increasing by ξ^2 with each technology generation [27]. Figure 1.5 shows the history of technology scaling and its effect on power density, indicating that we are quickly approaching the power density of a nuclear reactor. Note also that although power density is measured in W/cm^2, this is an approximation recognizing that most of the heat generated in a transistor is generated within a narrow depth (~100 nm [28]) while the chip may be over a millimeter thick.

1.2   Global and Local Temperature Variation

There are two types of temperature variation that affect system performance: global temperature variations and local temperature variations. Global temperature variations are caused by changes in ambient temperature or changes in cooling capacity. The United States military IC requirements for ambient temperature extend from −55°C to 125°C [29]. Increasing global chip temperatures will cause path latencies to exceed clock periods, resulting in functional failure.

The disparity in power dissipation between active units and inactive units can result in severe hot spots on a chip, creating large temperature variations which can reduce functionality or cause timing failure. Local chip temperature variations (also called intra-die temperature variations) can also result in communication errors between units with a large temperature differential. Intra-die temperature variations exceeding 50°C have been reported [30], as shown by the thermal map in Fig. 1.6, which shows the temperature gradient between a microprocessor core and an on-chip cache. Adaptive systems with temperature sensors ensure functionality over this wide range of conditions by adjusting cooling systems, supply voltage and/or operating frequency [31–33].

In addition to the risk of functional failure caused by delay increases, extended exposure to high temperatures can result in premature aging and electromigration (high temperatures result in increased electron energies, and

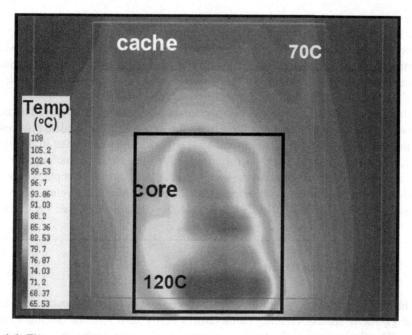

Fig. 1.6 Thermal map of an integrated microprocessor highlighting on-chip temperature variation [Intel'03]

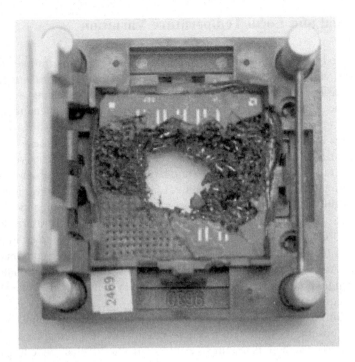

Fig. 1.7   Impact of thermal runaway on a test socket [35]

these high-energy particles can more easily damage the material lattice) additional failure mechanisms that affect chip performance. These are long-term issues (electromigration can take years to cause failures [14]), important for applications where there is limited or no access to a system once it has been deployed, such as space applications or implanted devices (e.g. pacemakers).

Temperature-related failure can also occur on a very short timescale. Thermal runaway is a serious problem resulting from the exponential dependence of temperature on subthreshold leakage current [34], I_{sub}, which is explained in detail in Chap. 2. Thermal runaway is the condition where an increase in temperature causes an increase in leakage current, and the increase in leakage current dissipates enough additional power to further increase the temperature, resulting in a cycle of increasing leakage and temperature that can have unstable consequences (see Fig. 1.7).

1.3   Thermal Control in VLSI Systems

The billions of devices in Very Large Scale Integration (VLSI) systems turn on and off at gigahertz speeds, creating temperatures easily sufficient for boiling water. Controlling the heat put out by these systems is becoming increasingly important to maintain reliability, increasingly expensive in terms of both design hours and cooling system costs, and increasingly complex to design and verify. Thermal design begins with a power budget—the maximum power dissipation of the chip. This number is heavily dependent on the application, which determines the cooling capacity and available power source. For example, IBM's high-end server chips will be placed in rooms with expensive cooling systems and specialized high-power wall sockets, resulting in the 8-core POWER 750 chip's power budget of 488 W [36]. In contrast, the Qualcomm dual-core Snapdragon mobile processor has no access to active cooling (there are no fans or vents in a smartphone), and also has a very limited battery energy source, resulting in a power budget of just 1.2 W [37].

This power budget determines the expected heat output of the chip, and designers have a number of options for controlling the associated thermal challenges that may arise. For example, server processors may utilize on-chip thermal control systems as well as off-chip cooling systems such as heat sinks, cooling fans, or even water-based cooling solutions; in contrast, chips with smaller power budgets (like a smart phone processor) do not need the same complex off-chip thermal management mechanisms. Despite the lack of off-chip thermal management, mobile systems may be required to operate over a wide range of ambient conditions, while a server-processor will likely sit in a carefully climate-controlled server farm. Thus, different application spaces have very different thermal requirements and thermal management systems.

Off-chip temperature-control was discussed in the previous section (see Fig. 1.4). On-chip mechanisms for thermal control depend on the chip's power budget. In applications with larger power densities, designers try to reduce the occurrence of hot spots to simplify the thermal control systems (hot spots may require individual thermal monitors, while a chip with an evenly distributed temperature may only need a single sensor). Hotspots can be reduced at the design stage using thermally-aware floor planning techniques [38], in which units with large power dissipations are spread evenly around a chip. Static timing analysis tools must also be aware of potential temperature differences between two locations on a chip to avoid communication errors [39]. At runtime, adaptive measures such as thermal throttling are commonly used [32], which improve reliability by adjusting the processor frequency and supply voltage when high-temperature conditions are detected. These adaptive systems can be useful for mobile applications as well, detecting when a chip is brought into a hot environment and reducing its processor speed to avoid timing failures. Some applications have very specific thermal requirements, which may call for alternative solutions such as different material systems. For example, SiC is extremely desirable for high-temperature ($\sim$600°C) sensor applications where conventional Si-based sensors would fail [40], such as engine exhaust systems.

Some thermal solutions treat local and global temperature variations differently. For example, floorplanning is only useful for managing local temperature variations, while turning on a cooling fan is only useful for managing the global chip temperature. Other approaches simplify the problem by handling temperature variations in the same manner regardless of whether they are global or local and regardless of whether they are on-chip or environmental variations.

In this book, we provide a number of different solutions to thermal issues. Some, such as the proposed low-power safety mode for multi-core chips in Chap. 4, react to local changes in temperature, while others such as the proposed Programmable Temperature Compensation Device method presented in Chap. 5 do not make a distinction between local and global temperature variations, providing uniform delay performance over the entire military-specified temperature range.

1.4 Book Overview

This book examines how nanoscale circuits are affected by the massive amounts of heat they generate. We present methods for detecting these effects or avoiding them altogether. We present a number of problems created by on-chip and environmental temperature variations, and propose solutions to these problems along with the trade-offs that need to be considered when implementing each solution.

Some solutions to scaling challenges will bring about even larger temperature problems, such as the increasing adoption of Silicon-on-Insulator (SoI) technologies. The compilation of SoI wafers into three-dimensional stacked wafer systems [41] will further increase power density (in cm^3 rather than cm^2) and reduce access to cooling (internal layers will have no contact with cooling surfaces), making temperature a truly critical concern for future system design.

To avoid erroneous computation resulting from increasingly significant variations in temperature, system control units must have detailed knowledge of on-chip temperature profiles. The creation of on-chip thermal maps such as those shown in Fig. 1.6 requires low-overhead, energy-efficient sensors that can be replicated in multiple regions of a chip. Normal and reverse temperature variation regimes (described in detail in Chap. 2) require sensors capable of detecting both the temperature of a unit and its temperature dependence. Chapter 2 also presents a more detailed background of how temperature affects material properties and circuit functionality, as well as the impact of scaling on temperature variation (including the use of high-κ dielectrics and metal gates). Our work on temperature sensing is presented in Chap. 3, including runtime variation-aware systems capable of detecting and reacting to overheating temperatures, a low-overhead temperature sensor capable of adaptively trading off energy for increased resolution, and a sensor to detect both normal and reverse temperature dependences.

Sensing alone is sufficient for applications where temperature is only tracked and recorded; however, the increase in power density and thermal variations in VLSI systems requires active temperature management systems—if the temperature approaches levels that could result in system failure, the system must adapt based on the sensor input to adjust parameters and ensure reliable operation. The temperature management systems can enable energy efficient adaptations, using only as much energy as necessary to complete a task and shutting down or scaling back performance when units are not being used. These principles have brought about techniques such as clock gating, power gating, adaptive voltage scaling

(AVS), adaptive body bias (ABB), and many others. AVS systems have proven to be a particularly excellent method for improving energy efficiency, and are currently in use in every major microprocessor, including Intel's Speedstep [42], AMD's PowerNow! [43], and IBM's EnergyScale [44] technologies.

AVS systems are described in detail in Chap. 4, where we examine the challenges of ensuring reliability to process, voltage, and temperature (PVT) variations in these systems which may have multiple operating points with very different temperature variation profiles.

Although adaptive systems can be designed to react to a wide variety of temperature conditions, some applications require stricter control of the impact of temperature on system parameters. For example, intra-die temperature variations can result in large clock skews that limit performance; also, voltage sensor data may be corrupted by changes in temperature, requiring tight bounds to achieve an accuracy target. In Chap. 5, we examine the concept of temperature-insensitive system design, comparing prior approaches and proposing a new approach that achieves the most accurate and most versatile temperature insensitivity to date.

Thus far we have assumed that temperature-related effects have a negative impact on system performance. In Chap. 6 we show that this is not always the case; under certain conditions we can exploit differences in the temperature dependence between neighboring units to improve overall performance. For example, interconnect links using low voltages often interface with transceivers operating at higher voltages. In this case, the low voltage links in the reverse temperature dependence regime are fastest at high temperatures, while the higher voltage transceivers are slower at high temperatures. A simple delay-borrowing mechanism allows us to average the two to improve performance compared to worst-case design.

Finally, we must recognize that regardless of our efforts to control and adapt to changes in temperature, thermally-accelerated aging mechanisms and thermal faults can still limit our system functionality. Intermittent and permanent faults resulting from variations and noise require variation-tolerant systems capable of detecting and correcting errors. Chapter 7 presents a unique runtime system for detecting and correcting transient, intermittent, and permanent errors in on-chip interconnect.

Future work and open problems in the areas we have discussed are presented in Chap. 8.

References

1. Middleton WEK (1966) The history of the thermometer and its use in meteorology. Johns Hopkins University Press, Baltimore
2. Behar MF (1932) Temperature and humidity measurement and control. Instruments Publishing Company, 113–122
3. Fludd R (1638) Philosophia Moysaica Goudae, 2
4. Tuoriniemi JT, Knuuttila TA (2000) Nuclear cooling and spin properties of Rhodium down to picoKelvin temperatures. Physica B Condensed Matter 280:474–478

5. Goren-Inbar N (2008) Fire out of Africa: a key to the migration of prehistoric man, says Hebrew University archaeological researcher. Hebrew University of Jerusalem, press release.
6. Thomas I (1941) Greek mathematical works, vol. II. Cambridge
7. Parigi G (1599) Archimedes' burning glass warfare. Painting
8. Gilbert W (1600) De magnete. Peter Short, London
9. Davy H, Davy J (1840) The collected works of Sir Humphry Davy. Smith Elder and Co, London
10. Seebeck TJ (1822) Magnetische polarization der metalle und erze durch temperature-differenzen. Akad Berlin Abh 265–374
11. Peltier JCA (1834) Investigation of the heat developed by electric currents in homogeneous materials and at the junction of two different conductors. Ann Chim Phys 56:371
12. Gould CA, Shammas NYA, Grainger S, Taylor I (2008) A comprehensive reviewer of thermoelectric technology, micro-electrical and power generation properties. 26th Int Conf in Microelectronics, 329–332
13. Faraday M (1839) Experimental researches in electricity, 1st ed. Bernard Quaritch, London
14. Bernstein K et al (2006) High-performance CMOS variability in the 65-nm regime and beyond. IBM J Res And Dev 50:433–449
15. Frank DJ et al (2001) Device scaling limits of Si MOSFETs and their application dependencies. Proc IEEE 89:259–288
16. Frank DJ (2002) Power-constrained CMOS scaling limits. IBM J Res and Dev 46:235–244
17. Semiconductor Industry Association (2007) International technology roadmap for semiconductors, executive summary. [Online] http://www.itrs.net
18. Rabaey J, Chandrakasan A, Nikolic B (2003) Digital integrated circuits: a design perspective, 2nd ed. Prentice Hall, New Jersey
19. Rusu S, Sachdev M, Svensson C, Nauta B (2002) Trends and challenges in VLSI technology scaling towards 100 nm. 7th Asia and South Pacific Design Automation Conf 16–17
20. Goodson KE, Flik MI, Su LT, Antoniadis DA (1995) Prediction and measurement of temperature fields in silicon-on-insulator electronic circuits. J Heat Transfer 117:574–581
21. Su LT, Goodson KE, Antoniadis DA, Flik MI, Chung JE (1992) Measurement and modeling of self-heating effects in SOI nMOSFETs. Int Electron Devices Mtg 13–16
22. Wei J (2008) Challenges in cooling design of CPU packages for high-performance servers. Heat Transfer Engineering 29:178–187
23. Intel (2008) Intel Core2 Duo processor E8000 and E7000 series datasheet. [Online] http://download.intel.com/design/processor/datashts/318732.pdf
24. Intel (2008) Intel Core2 Extreme processor QX9000 series and Intel Core2 Quad processor Q9000 and Q8000 series datasheet. [Online] http://download.intel.com/design/processor/datashts/318726.pdf
25. Pollack F (1999) New microprocessor challenges in the coming generations of CMOS technologies. 32nd Ann ACM/IEEE Int Symp on Microarchitecture 2
26. Intel (2005) Intel Pentium 4 Processor 6xx Sequence and Intel Pentium 4 Processor Extreme Edition datasheet. [Online] http://download.intel.com/design/processor/datashts/318726.pdf
27. Wong HP, Frank DJ, Solomon PM, Wann CHJ, Welser JJ (1999) Nanoscale CMOS. Proc IEEE 47:537–570
28. Lin CL, Yeh WK (2011) Impact of SOI thickness on device performance and gate oxide reliability of Ni fully silicide metal-gate strained SOI MOSFET. Microelectronic Engineering 88:228–234
29. US Dept of Defense (2007) Integrated circuits (microcircuits) manufacturing, general specification, std MIL-PRF-38535H. Washington DC
30. Sato T, Ichimiya J, Ono N, Hachiya K, Hashimoto M (2005) On-chip thermal gradient analysis and temperature flattening for SoC design. IEICE Trans Fundamentals E88-A: 3382–3389
31. Dinh JS, Korinsky GK (1996) Temperature dependent fan control circuit for personal computer. US Patent 5526289

32. Tschanz J et al (2007) Adaptive frequency and biasing techniques for tolerance to dynamic temperature-voltage variations and aging. IEEE Int Solid-State Circuits Conf 292–604
33. Elgebaly M, Sachdev M (2007) Variation-aware adaptive voltage scaling system. IEEE Trans Very Large Scale Integr Syst 15:560–571
34. Narendra SG, Chandrakasan AP (2005) Leakage in nanometer CMOS technologies. Springer: USA
35. Vassighi A, Semenov O, Sachdev M, Keshavarzi A, Hawkins C (2004) CMOS IC technology scaling and its impact on burn-in. IEEE Trans Device and Materials Reliability 4:208–221
36. IBM (2010) Power your planet: Smarter systems for a smarter planet. Presentation
37. Gwennap L (2010) Two-headed snapdragon takes flight: Qualcomm samples dual-CPU mobile processor at 1.2 GHz. The Linley Group
38. Yu H, Hu Y, Liu C, He L (2007) Minimal skew clock embedding considering time variant temperature gradient. Int Symp Physical Design 173–180
39. Dasdan A, Hom I (2006) Handling inverted temperature dependence in static timing analysis. ACM Trans Design Automation of Electronic Syst 11:306–324
40. Nakagomi S et al (1997) Influence of carbon monoxide, water and oxygen on temperature catalytic metal-oxide-silicon carbide structures. Sensors and Actuators B 45:183–191
41. Pavlidis VF, Friedman EG (2008) Three-dimensional integrated circuit design. Morgan Kaufmann:MA
42. Gochman S et al (2003) The Intel Pentium M processor: microarchitecture and performance. Intel Tech J 7:21–38
43. Advanced Micro Devices (2002) AMD PowerNow! technology brief. [Online] http://www.amd.com/us-en/assets/content_type/DownloadableAssets/Power_Now2.pdf
44. McCreary HY et al (2007) EnergyScale for IBM Power6 microprocessor-based systems. IBM J Res and Dev 51:775–786

Chapter 2
Temperature Effects in Semiconductors

The changes in temperature described in the previous chapter affect the speed, power, and reliability of our systems. Throughout this book, we will examine all three of these metrics, though the majority of our discussion will be on how temperature affects the speed performance. In this chapter, we discuss the problem of temperature variation at the device and circuit level. In Sect. 2.1, we provide a background on the material dependences on temperature. In Sect. 2.2, the normal and reverse temperature dependence regimes are described. In Sect. 2.3, we explore how these dependences change with technology scaling and the introduction of new processing materials, such as high-κ dielectrics and metal gates.

2.1 Material Temperature Dependences

In this section we provide details about the impact of temperature on the MOSFET energy band gap, carrier density, mobility, carrier diffusion, velocity saturation, current density, threshold voltage, leakage current, interconnect resistance, and electromigration.

2.1.1 Energy Band Gap

Temperature affects the properties of electronic systems in a number of fundamental ways. The most fundamental of properties is the energy band gap, E_g, which is affected by temperature according to the Varshni equation [1]

$$E_g(T) = E_g(0) - \frac{\alpha_E T^2}{T + \beta_E} \tag{2.1}$$

where $E_g(0)$ is the band gap energy at absolute zero on the Kelvin scale in the given material, and α_E and β_E are material-specific constants. Table 2.1 [2] provides these

D. Wolpert and P. Ampadu, *Managing Temperature Effects in Nanoscale Adaptive Systems*, DOI 10.1007/978-1-4614-0748-5_2, © Springer Science+Business Media, LLC 2012

Table 2.1 Varshni equation constants for GaAs, Si, and Ge [2]

Material	$E_g(0)$ (eV)	α_E (eV/K)	β_E (K)
GaAs	1.519	$5.41*10^{-4}$	204
Si	1.170	$4.73*10^{-4}$	636
Ge	0.7437	$4.77*10^{-4}$	235

Fig. 2.1 Energy band gap temperature dependence of GaAs, Si, and Ge

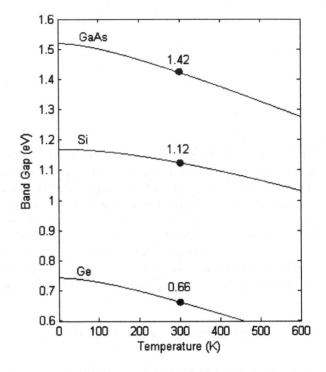

constants for three material structures. Table 2.1 and (2.1) are used to generate Fig. 2.1, which shows how the band gaps of the three materials decrease as temperature increases (the labeled points are the band gap of each material at room temperature).

2.1.2 Carrier Density

Carrier densities affect electrical and thermal conductivity, and are a function of the effective density of states in the appropriate band (conduction for n-type, valence for p-type), the Fermi energy level in the material (which is a function of temperature and dopant concentrations), and the temperature as given by the following equations:

$$n = N_C e^{-\frac{E_C - E_F}{kT}}$$

(2.2)

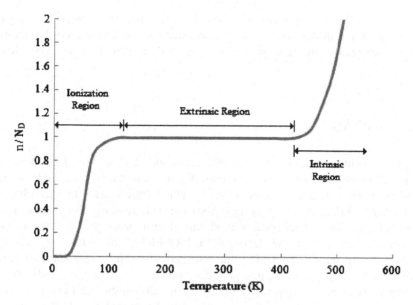

Fig. 2.2   Temperature dependence of n in a doped semiconductor

$$p = N_V e^{-\frac{E_F - E_V}{kT}} \tag{2.3}$$

where n is the electron density, p is the hole density, N_C is the density of states in the conduction band, N_V is the density of states in the valence band, E_C is the conduction band energy level, E_V is the valence band energy level, E_F is the Fermi energy level, $k = 1.38 \cdot 10^{-23}$ J/K is the Boltzmann constant, and T is temperature.

The temperature dependence of carrier density is shown in Fig. 2.2 for a doped material. In the ionization region, there is only enough latent energy in the material to push a few of the dopant carriers into the conduction band. In the extrinsic region, which is the desired region of operation, the carrier concentration is flat over a wide range of temperatures; in this region, all of the dopant carriers have been energized into the conduction band (i.e. $n \approx N_D$) and there is very little thermal generation of additional carriers. As the temperature increases, the extrinsic region turns into the intrinsic region, and the number of thermally generated carriers exceeds the number of donor carriers. The intrinsic carrier concentration in a material n_i is generally much smaller than the dopant carrier concentration at room temperature, but n_i ($=n \cdot p$) has a very strong temperature dependence [2]

$$n_i \propto T^{1.5} e^{-\frac{E_{g0}}{2kT}} \tag{2.4}$$

where E_{g0} is the energy band gap at $T = 0$ K. Depending upon the dopant concentration, the number of thermally generated carriers can exceed the number of dopant-generated carriers, increasing the potential for thermal variation problems.

2.1.3 Mobility

We pay particular attention to the temperature and electric field dependence of mobility, as mobility is one of the two main factors (the other is threshold voltage) resulting in the MOSFET temperature behavior shown later in this chapter. The carrier mobility, μ (cm^2/V·s), describes the drift velocity of a particle in an applied electric field. Under small to moderate electric fields, $\mu = v_d/\xi$ where v_d is the drift velocity, and ξ is the electric field. MOSFET mobility has very complex temperature dependence, defined by the interplay of the following four scattering parameters: phonon scattering μ_{ph}, surface roughness scattering μ_{sr}, bulk charge Coulombic scattering μ_{cb}, and interface charge Coulombic scattering μ_{int} [3]. Each of these scattering parameters is related to the temperature of the material, T, and the effective transverse electric field in the channel, ξ_{eff}, which is approximated as [4, 5]

$$\xi_{eff} \approx \frac{(\eta Q_{inv} + Q_b)}{\varepsilon_{Si}} \approx \frac{(V_{gs} + V_T)}{6 T_{ox}} \tag{2.5}$$

where η is a constant ($\eta \approx 0.4$ in PMOS devices and $\eta \approx 0.5$ in NMOS devices), Q_{inv} is the inversion layer charge density, Q_b is the substrate depletion charge density, and $\varepsilon_{Si} = 11.7$ is the relative permittivity of Silicon. This approximation is not very convenient for circuit analysis, so ξ_{eff} is also approximated in terms of the gate-source voltage V_{gs}, the threshold voltage V_T, and gate oxide thickness T_{ox}.

The Berkeley Short-Channel IGFET Model (BSIM), one of the most widely used simulation models, combines these four scattering parameters into an effective mobility, μ_{eff} [3] using Matthiessen's rule

$$\frac{1}{\mu_{eff}(T, E_{eff})} \propto \frac{1}{\mu_{ph}(T, E_{eff})} + \frac{1}{\mu_{sr}(T, E_{eff})} + \frac{1}{\mu_{cb}(T, E_{eff})} + \frac{1}{\mu_{int}(T, E_{eff})} \tag{2.6}$$

Phonon scattering refers to the potential for an electron to be scattered by a lattice vibration. As temperature increases, lattice vibrations increase and the probability of an electron being scattered by the lattice increases; thus, high temperature mobilities are limited by phonon scattering ($\mu_{ph} \propto T^{-3/2}$), causing mobility to decrease as temperature increases as shown in Fig. 2.3a. Surface roughness scattering becomes dominant when high electric fields pull electrons closer to the Si/SiO$_2$ surface ($\mu_{sr} \propto \xi_{eff}^{-2.1}$).

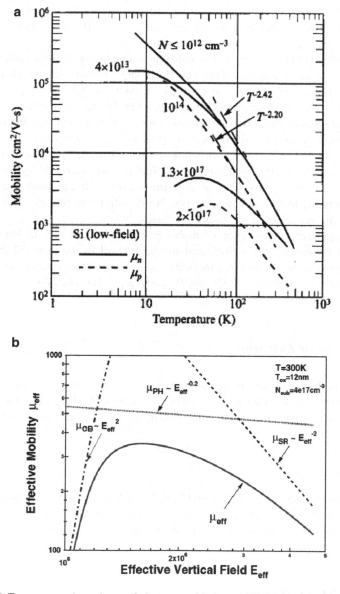

Fig. 2.3 (a) Temperature dependence of electron and hole mobilities in Si for different dopant concentrations [2], (b) Field dependence of mobility [7]

At low temperatures, electrons move more slowly, and lattice vibrations are small as well; thus, the ion impurity forces which have little impact on high-energy particles become the dominant limit to mobility. In this regime, decreasing temperature extends the amount of time electrons spend passing an impurity ion, causing mobility to decrease as temperature decreases ($\mu_{cb} \propto T$). This effect is emphasized in the high dopant concentration curves shown in Fig. 2.3a, where mobility decreases with decreasing temperature (e.g. the $\mu_n = 1.3 \cdot 10^{17}$ dopant concentration line below ~30 K).

The electric field dependence of mobility is shown in Fig. 2.3b. In bulk Coulombic scattering, increasing ξ_{eff} increases the charge density in the channel; the associated charge screening reduces the impact of μ_{cb} ($\propto \xi_{eff}^2$). At low temperatures, the interface charges have two conflicting dependences. Reduced temperature reduces the carriers' thermal velocity, which increases the impact of interface charges; however, the reduced thermal velocity also reduces the screening effect [6], and this reduction in screening dominates the temperature dependence ($\mu_{int} \propto T^{-1}$). The electric field screening effect is also weakened by the reduced thermal velocity ($\mu_{int} \propto \xi_{eff}$, not ξ_{eff}^2 as in the μ_{cb} limit). In this book, we consider devices operating in the phonon scattering limit, with temperatures >200 K; thus, mobility will decrease as temperature increases.

The temperature dependence of mobility plays a major role in temperature-aware system design, and is discussed in more detail in the next subsection. In room temperature Si, the electron mobility, μ_n, is nearly three times as large as the hole mobility, μ_p, with $\mu_n = 1,350$ cm^2/V·s and $\mu_p = 480$ cm^2/V·s.

2.1.4 Carrier Diffusion

Diffusion is the movement of particles from a region of high concentration to a region of low concentration. Carrier diffusion coefficients D_n and D_p (for electrons and holes, respectively) are related to mobility by the Einstein relationship

$$\frac{D}{\mu} = \frac{kT}{q} \qquad (2.7)$$

Here, q is the charge on an electron ($1.6 \cdot 10^{-19}$ C), and kT/q is an important value known as the thermal voltage, ϕ_T. At room temperature (300 K), $\phi_T = 0.0259$ V. D_n and D_p in room temperature silicon are 36 and 12 cm^2/s, respectively.

2.1.5 Velocity Saturation

Although saturation velocity has been recently found to be a dominant temperature-dependent parameter, notable work had been performed in this area as far back as 1970 [8] using device lengths of 10 μm. In the BSIM4 device model, the impact of temperature on velocity saturation v_{sat} is modeled by [9]

$$v_{sat} = v_{sat0} \cdot [1 - \alpha_{v_{sat}}(T - T_0)] \qquad (2.8)$$

where v_{sat0} is the saturation velocity at nominal temperature (T_0) and α_{vsat} is the saturation velocity temperature coefficient. Qualitatively, velocity saturation is the point at which increases in energy no longer cause carrier velocity to increase; instead, the additional energy is lost to phonon generation through lattice interactions.

In the results presented in this book, devices operate in the velocity saturation regime; thus, the impact of temperature on saturation velocity (increasing temperature decreases v_{sat}) is one of the most important criteria affecting the overall impact of temperature on device current, as will be shown later in this chapter.

2.1.6 Current Density

The temperature dependence of the carrier concentrations, mobilities and diffusion coefficients affect the temperature behavior of the carrier current densities, with the carrier densities defined by the following formulas [10]:

$$J_N = q\mu_n n \xi + qD_n \nabla n \tag{2.9}$$

$$J_P = q\mu_p p \xi - qD_p \nabla p \tag{2.10}$$

where J_N and J_P are the electron and hole current densities, respectively. The first term in each equation is the *drift* component of the total current, with μ_n and μ_p corresponding to the electron and hole mobilities, respectively; ξ is the electric field. The second term in each equation is the *diffusion* component of the total current, with ∇n and ∇p corresponding to the electron and hole concentration gradients (if there is no concentration gradient, there is no diffusion). The temperature dependent parameter in the second term is the diffusion coefficient. Increased temperature increases particle kinetic energy, increasing the diffusion component of total current. The drift component of the total current has two temperature dependent parameters, the mobility and the carrier density. The mobility term was shown in Fig. 2.3 to decrease as temperature increases (in the lattice vibration-limited case) while the carrier density remains nearly fixed with temperature over the extrinsic range (our intended range of operation), as indicated by Fig. 2.2. Thus, we determine that the drift component of the total current decreases as temperature increases.

The drift and diffusion currents have opposing temperature dependencies, which causes the net current change to depend on the applied electric field. In the high-field (drift-dominated) case, current decreases as temperature increases; in the low-field (diffusion-dominated) case, current increases as temperature increases. However, if the system in question is a multi-voltage system, and the system has both drift- *and* diffusion-dominated components, the impact of temperature variation may become less well-defined. The difference between a drift-dominated system and a diffusion-dominated system is defined by the threshold voltage, V_T. We will show that the temperature dependences of mobility and threshold voltage result in some very interesting device behavior.

2.1.7 Threshold Voltage

The MOSFET threshold voltage is given by [2]

$$V_T = V_{FB} + 2\phi_F + \gamma\sqrt{2\phi_F} \tag{2.11}$$

where $V_{FB} = \phi_{gs} - (Q_{ss}/C_{ox})$ is the flat band voltage, with the gate-substrate contact potential $\phi_{gs} = \phi_T \ln(N_A N_G/n_i^2)$, N_A and N_G are the substrate and gate doping concentrations, respectively, Q_{ss} the surface charge density, and C_{ox} the oxide capacitance; $\gamma = C_{ox}(2q\varepsilon_{Si}N_A)^{0.5}$ is a body effect parameter, with ε_{Si} the relative permittivity of Si; $\phi_F = \phi_T \ln(N_A/n_i)$ is the Fermi energy with the thermal voltage $\phi_T = kT/q$, and n_i the intrinsic carrier concentration of Si.

Of the parameters in (2.11), ϕ_{gs} and ϕ_F vary with temperature (each contains ϕ_T and n_i terms). The threshold voltage temperature dependence $\partial V_T/\partial T$ may thus be written as [11]

$$\frac{\partial V_T}{\partial T} = \frac{\partial \phi_{gs}}{\partial T} + 2\frac{\partial \phi_F}{\partial T} + \frac{\gamma}{\sqrt{2\phi_F}}\frac{\partial \phi_F}{\partial T} \tag{2.12}$$

where the temperature dependencies of ϕ_{gs} and ϕ_F are [11]

$$\frac{\partial \phi_{gs}}{\partial T} = \frac{1}{T}\left(\phi_{gs} + \frac{E_{G0}}{q} + \frac{3kT}{q}\right) \tag{2.13}$$

$$\frac{\partial \phi_F}{\partial T} = \frac{1}{T}\left[\phi_F - \left(\frac{E_{G0}}{2q} + \frac{3kT}{2q}\right)\right] \tag{2.14}$$

Filanovsky [11] used empirical parameters from a 0.35 μm CMOS technology to determine that the three terms in (2.12) are -3.1, 2.7, and -0.43 mV/K, resulting in a net threshold temperature coefficient of -0.83 mV/K. The threshold voltage in a MOSFET is commonly modeled to decrease linearly with increasing temperature; the parameter is plotted in Fig. 2.4 over a range of oxide thicknesses d and dopant concentrations N_A.

2.1.8 Leakage Current

Subthreshold leakage current I_{sub} is exponentially dependent on temperature, as shown in Fig. 2.5; a common rule of thumb is that leakage current doubles for every 10°C increase in temperature [12]. When $V_{GS} = 0$, I_{sub} may be represented by the Shockley diode model

$$I_{sub} = I_0\left(e^{\frac{V_{DS}}{\phi_T}} - 1\right) \tag{2.15}$$

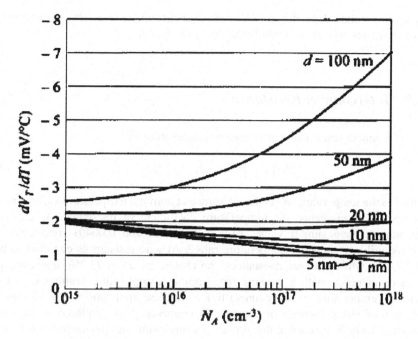

Fig. 2.4 Change in threshold voltage temperature dependence at room temperature vs. dopant concentration, with oxide thickness d [2]

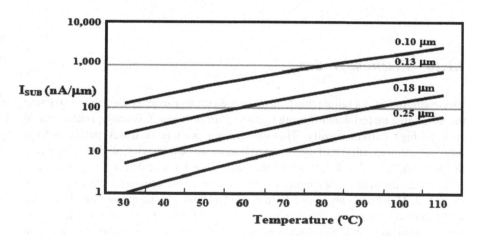

Fig. 2.5 Temperature dependence of subthreshold leakage current ($V_{GS} = 0$ V) [14]

$$I_0 = ATe^{-\frac{1.12}{2\phi_T}} \qquad\qquad (2.16)$$

where I_0 is the reverse saturation current [12], A is a constant, and V_{DS} is the drain-source voltage. Recalling that $\phi_T = kT/q$, we see that I_0 is responsible for the exponential temperature dependence shown in Fig. 2.5.

The temperature dependence of gate leakage current has been shown to be very minor compared to that of subthreshold leakage current [13].

2.1.9 Interconnect Resistance

The interconnect resistance R is related to temperature by

$$R(T) = R_0[1 + \alpha_R(T - T_0)] \tag{2.17}$$

where T is the temperature, R_0 is the resistance at nominal temperature T_0, and α_R is an empirical term named the temperature coefficient of resistance. Al and Cu interconnects have similar values of α_R—0.004308 and 0.00401, respectively. Over the military-specified temperature range, Al wire resistances can change by up to 77.5% while Cu wire resistances can change by up to 72.2%. Interconnect resistance increases with increasing temperature, complicating evaluation of the impact of temperature on interconnect links—in these applications, the MOSFET currents may either increase or decrease in temperature (as explored in the next subsection), which means that the impact of temperature on interconnect resistance can either add to the system temperature dependence or reduce the temperature dependence, depending on the operating conditions.

2.1.10 Electromigration

Electromigration is a failure mechanism caused by high-energy electrons impacting the atoms in a material and causing them to shift position. It is most problematic in areas of high current density. This can form a positive feedback path can form where electromigration will cause an atom to move down a wire, slightly narrowing the wire width at that location and increasing the current density; this increased current density then further increases electromigration, causing more atoms to be displaced. This brings about two failure mechanisms: (1) the narrowing of the wire will increase wire resistance, which may cause a timing failure if a signal can no longer propagate within the clock period, or (2) electromigration will continue until the wire completely breaks, allowing no further current flow and resulting in functional failure.

Electromigration's impact on a system's reliability is measured in terms of a mean time to failure (MTTF) using Black's equation [15]

$$MTTF = A_j \cdot J^{-n_j} \cdot e^{\frac{E_a}{kT}} \tag{2.18}$$

where A_j is a constant related to the cross-sectional area of a wire, J is the current density, n is a constant scaling factor, E_a is the activation energy, k is the Boltzmann constant, and T is temperature. Thus, the MTTF is exponentially dependent on temperature.

2.2   Normal and Reverse Temperature Dependence

Changes in temperature affect system speed, power, and reliability by altering the threshold voltage [11], mobility [11], and saturation velocity [16] in each device. The resulting changes in device current can lead to failures in timing, cause systems to exceed power or energy budgets, and result in communication errors between IP cores. The temperature relationships for MOSFET mobility, threshold voltage, and velocity saturation are related to temperature using the following empirical expressions [17]:

$$\mu(T) = \mu_0 (T/T_0)^{\alpha_\mu} \tag{2.19}$$

$$V_T(T) = V_{T0} + \alpha_{V_T}(T - T_0) \tag{2.20}$$

$$v_{sat}(T) = v_{sat0} + \alpha_{v_{sat}}(T - T_0) \tag{2.21}$$

where T is the temperature; T_0 is the nominal temperature; μ_0, V_{T0}, and v_{sat0} are the mobility, threshold voltage, and saturation velocity at T_0, respectively; α_μ, α_{VT}, and α_{vsat} are empirical parameters named the mobility temperature exponent, threshold voltage temperature coefficient, and saturation velocity temperature coefficient, respectively, where $\alpha_\mu \approx -1.3$, $\alpha_{VT} \approx -3$ mV/°C, and $\alpha_{vsat} \approx -97$ m/(s·°C). Two temperature dependencies exist: the normal dependence (ND) region, where drain current (I_D) *decreases* with increasing temperature, and the reverse dependence (RD) region, where I_D *increases* with increasing temperature [18]. Between the two regions, there is a supply voltage where the impact of temperature on delay is minimized. This is referred to as the temperature-insensitive voltage V_{INS} [19], and as technology scales this voltage approaches nominal voltage.

In the temperature region of concern (between −55°C and 125°C, the range of military operating temperatures [20]), μ, V_T, and v_{sat} all decrease with increasing temperature. Examining the velocity-saturated MOSFET drain current $I_D(T)$ [21] we see that decreasing v_{sat} *decreases* I_D, while decreasing V_T *increases* I_D [22].

$$I_D(T) = v_{sat}(T) \cdot W \cdot P_s \cdot [V_{GS} - V_T(T)]^\alpha \tag{2.22}$$

Where W is the device width, P_s is a technology-specific constant, V_{GS} is the MOSFET gate-source voltage, and α is a technology-specific exponent. The temperature dependence of the MOSFET drain current, dI_D/dT, can be determined by the sum of the impacts of v_{sat} and V_T on I_D, composed of four values—(1) the change in velocity saturation for a change in temperature, dv_{sat}/dT, (2) the change in threshold voltage for a change in temperature, dV_T/dT, (3) the change in device current for a change in velocity saturation, $\partial I_D/\partial v_{sat}$, and (4) the change in device current for a change in threshold voltage, $\partial I_D/\partial V_T$:

$$\left.\frac{dI_D}{dT}\right|_{Tot} = \left.\frac{dI_D}{dT}\right|_{v_{sat}} + \left.\frac{dI_D}{dT}\right|_{v_T} = \frac{\partial I_D}{\partial v_{sat}} \cdot \frac{dv_{sat}}{dT} + \frac{\partial I_D}{\partial v_T} \cdot \frac{dv_T}{dT} \qquad (2.23)$$

$dI_D/dT|_{vsat}$ is negative, and $dI_D/dT|_{VT}$ is positive. At nominal voltage in conventional CMOS technologies, the magnitude of $dI_D/dT|_{vsat}$ is greater than the magnitude of $dI_D/dT|_{VT}$; thus, circuits at nominal voltages become slower as temperature increases. However, as V_{GS} approaches V_T, a change in V_T has a larger impact on I_D; thus, at lower supply voltages, the magnitude of $dI_D/dT|_{vsat}$ is less than the magnitude of $dI_D/dT|_{VT}$, and device delay decreases as temperature increases (the reverse temperature dependence). V_{INS} occurs where $dI_D/dT|_{Tot}$ approaches zero, with $dI_D/dT|_{vsat} \approx -dI_D/dT|_{VT}$; however, because v_{sat} and V_T differ between NMOS and PMOS devices, each type of device has a different value of V_{INS}. The dependence regions are shown in Fig. 2.6 for plots of the current through diode-connected PMOS and NMOS devices in a 90 nm technology model [23] over the range of military operating temperatures. In Fig. 2.6a, V_{INS} occurs in the shaded regions, with higher voltages exhibiting the normal temperature dependence and lower voltages exhibiting the reverse temperature dependence.

The reverse temperature dependence is occasionally referred to as temperature inversion, while the normal and reverse temperature dependences are also referred to as negative (for normal dependence) and positive (for reverse dependence) current-temperature (I-T) slopes. In this document, we will use the $\pm$ I-T slope terminology as shorthand for the normal and reverse temperature dependences.

The difference between the 125°C and −55°C endpoints of Fig. 2.6a is presented in Fig. 2.6b. In Fig. 2.6b, V_{INS} is indicated in each device by the minimum points in each curve; the absolute minimum for a 1:1 sizing ratio occurs at 345 mV, corresponding to an 18% total change in current over the entire 180°C range of ambient temperatures.

2.2.1 Discovery of the Normal and Reverse Temperature Dependences

This book is by no means the first document to report on the reverse temperature dependence. Indeed, what we name the reverse temperature dependence (i.e. the increasing of electrical conduction with increasing temperature) was first discovered by Faraday with his silver sulphide experiments mentioned in the previous chapter. However, the mechanism detected by Faraday was quite different than the

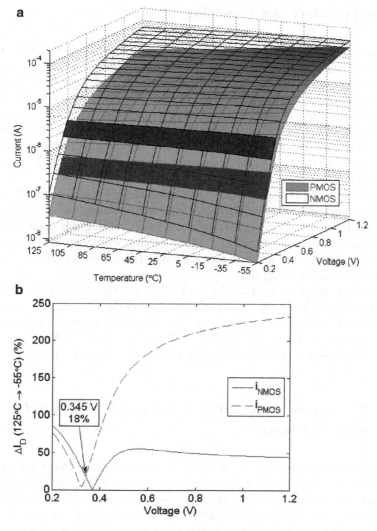

Fig. 2.6 (a) Device current across a range of temperatures and supply voltages in a 90 nm technology, (b) temperature change from 125°C to −55°C

mechanisms causing the normal and reverse temperature dependences in our modern Silicon electronics. The first recorded mention of the reversal of the temperature dependence describing the trade-off between mobility and threshold voltage is attributed to C. Park, *et al.*, from Motorola in 1995, in a conference paper exploring the impact of temperature on integrated circuits at very low voltages [18]. In the time since, the reversal of the temperature dependence has been explored in great detail [11, 24, 25], including magazines, patents, and journal papers, and is now being considered in industry-standard tools [26].

2.3 Temperature and Technology Scaling

V_{INS} occurs at increasingly larger supply voltages as technology is scaled, and is fast approaching the nominal supply voltage V_{NOM} (which is steadily decreasing as technology scales) as shown in Table 2.2, particularly with the introduction of high-κ dielectrics to replace SiO_2 [27] (high-κ dielectrics reduce μ and change $\partial\mu/\partial T$ [28], altering the balance of the μ and V_T impacts). The change in V_{INS} as technology scales is caused by changes in the threshold voltage and saturation velocity [29]. The data in Table 2.2 were generated using predictive technology models [23], with the 45, 32, and 22 nm data points using high-κ dielectric/metal gate models.

As V_{INS} approaches V_{NOM}, adaptive systems which vary supply voltage to reduce energy consumption or improve reliability will have operating voltages in both the normal and reverse temperature dependence regions, making it unclear if circuits will increase in speed or decrease in speed as chip temperatures increase, and exacerbating problems associated with inter-die temperature variation; solutions to this issue are discussed in Chap. 3.

The current dependencies for a 22 nm technology with high-κ dielectrics and metal gates are shown in Fig. 2.7. As shown, V_{INS} in the PMOS device increases from ~375 mV at 90 nm to ~575 mV at 22 nm, with the nominal supply voltage decreasing from 1.2 V at 90 nm to 0.8 V at 22 nm. The 22 nm NMOS device is in the reverse temperature dependence region even at the nominal supply voltage.

2.3.1 The Reverse Temperature Effect and High-κ/Metal Gate Technology

V_T, μ, v_{sat} and nominal supply voltage are all technology dependent parameters, with predicted values available down to the 22 nm node [23, 30]. Use of high-κ dielectrics and metal gates to alleviate nanoscale gate leakage problems also alters V_T, μ, and v_{sat} [31, 32]. The combination of these changes makes it difficult to determine the effect of temperature on device performance. Two dependences exist, as mentioned in the prior subsection: a normal temperature dependence, where current decreases as temperature increases, and a reverse temperature dependence [18, 19], where current increases as temperature increases.

Table 2.2 V_{INS} approaches V_{NOM} as technology scales

Technology (nm)	V_{NOM} (V)	V_{INS} (V)	V_{INS}/V_{NOM}
90	1.2	0.39	0.33
65	1.1	0.40	0.36
45	1.0	0.61	0.61
32	0.9	0.69	0.77
22	0.8	0.73	0.91

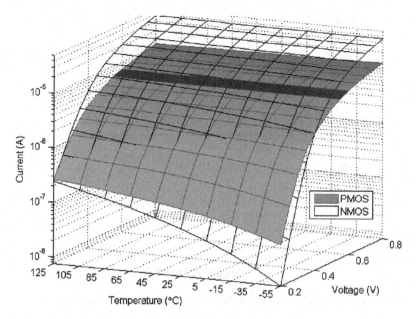

Fig. 2.7 Temperature dependence of device current across a range of supply voltages in a 22 nm high-κ/metal gate technology

These parameters are further complicated by environmental requirements (military specifications call for a range of -55–$125°C$ [20]) and intra-die temperature variation (shown to exceed $50°C$ [33]). To account for the wide range of conditions, as well as process and voltage variations, variation-tolerant adaptive systems have been used to guarantee functionality by adjusting operating voltages and frequencies [34–36]; however, these systems with multiple voltage modes make the above-mentioned temperature effects even more difficult to determine.

For large gate overdrives ($V_{GS}-V_T > V_{INS}$), the temperature dependence of a device is dominated by the dependence of v_{sat}, while for small gate overdrives ($V_{GS}-V_T < V_{INS}$), small changes in V_T can cause large changes in current, resulting in a temperature dependence dominated by V_T. Further examination of these effects in SiO_2 dielectric, polysilicon gate devices is available in [18, 19].

In nanoscale devices, high-κ dielectrics and metal gates have been introduced to reduce gate leakage due to thinning gate oxides and reduce the depletion effects of polysilicon gates [31, 32]; unfortunately, these techniques have the effect of dramatically altering the temperature dependence of I_D. The extent of the change is shown in Fig. 2.8, which compares 45 nm predictive technology models [23] of both SiO_2/poly gate (dashed line) and high-κ/metal gate (solid line) devices. Each line in Fig. 2.8 shows the change in delay of an inverter (sizing ratio $\beta = 2$) from $-55°C$ to $125°C$. For example, at 0.62 V, the high-κ/metal gate inverter delay is unchanged from $-55°C$ to $125°C$, resulting in the 0.62 V point occurring on the 0% line. This 0% intersect point on each curve represents V_{INS}. As shown, V_{INS} in the high-κ/metal gate is 40% higher than in the SiO_2/poly gate devices.

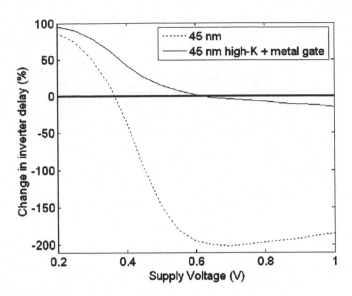

Fig. 2.8 Effect of high-κ dielectric and metal gate on temperature dependence

The normal dependence region is below the 0% line, and the reverse dependence region is above the 0% line.

Fig. 2.9a shows the change in PMOS device current from $-55°C$ to $125°C$ at the 45, 32, and 22 nm technology nodes (with nominal voltages of 1, 0.9, and 0.8 V, respectively). As shown, V_{INS} increases by ~40 mV per technology node, with V_{INS} at 22 nm equal to 0.56 V. The NMOS device response, shown in Fig. 2.9b, is in the reverse temperature dependence region over the entire range of operating voltages at the 32 and 22 nm nodes.

The PMOS and NMOS devices are combined into an inverter with $\beta = 2$ in Fig. 2.9c. As shown, V_{INS} approaches 90% of nominal voltage in the 22 nm node. As β increases, the stronger PMOS effect decreases V_{INS}. Thus, adaptive voltage systems may easily wind up straddling both temperature domains in nanoscale systems, making temperature-aware design increasingly critical as technology scales.

Reverse temperature dependence at near nominal voltages complicates variation-tolerant system design, which uses multiple supply voltages to adjust for changes in process, voltage, and temperature. The additional complexity needed to account for both normal and reverse temperature dependence depends on the available design time information. If the system can be fully characterized at design time, then the multiple dependences can be programmed into the voltage and frequency look-up table entries [34] to ensure that the system adapts in the correct direction given a change in temperature. For example, whereas a low-voltage system would generally reduce the frequency as temperature increases, in the reverse dependence region the system would have to reduce the frequency when temperature decreases.

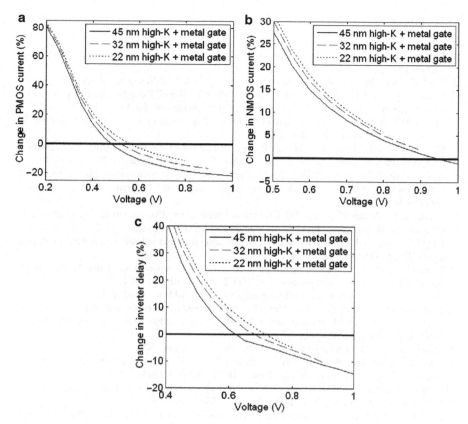

Fig. 2.9 Changes in (**a**) PMOS current, (**b**) NMOS current, and (**c**) inverter delay over the −55°C to 125°C temperature range

The reverse temperature effect is particularly important to consider in adaptive systems because of thermal runaway. In the normal dependence region, temperatures are prevented from increasing to dangerous levels because the delay becomes so large that the adaptive system is forced to reduce the clock frequency, reducing the energy and therefore the temperature. In the reverse temperature dependence region, circuits continue to speed up as temperature increases; there is no delay limit on high temperature operation. The higher temperatures could result in thermal runaway resulting from the exponential temperature dependence of leakage current [37], which may already be dominating the total power consumption in the nanoscale regime [38].

If the temperature dependences are not known at design time (from tool limitations, process variations, unknown IR drops, etc.), there are two options to ensure variation-tolerance. The system may be designed with large enough guardbands that it can operate correctly over the entire temperature range regardless of the temperature dependence, though this will result in a large reduction in delay performance. Another option is to use a temperature dependence sensor to determine the temperature dependence at each operating voltage; we propose the first temperature dependence sensor in Chap. 3.

References

1. Varshni YP (1967) Temperature dependence of the energy gap in semiconductors. Physica 34:149–154
2. Sze SM (1981) Physics of semiconductor devices, 2nd ed. John Wiley and Sons, NY
3. Chain K, Huang JH, Duster J, Ko PK, Hu C (1997) A MOSFET electron mobility model of wide temperature range (77–400K) for IC simulation. Semicond Sci Technol 12:355–358
4. Sabnis AG, Clemens JT (1979) Characterization of the electron mobility in the inverter Si surface. Int Electron Devices Mtg 18–21
5. Chen K, Wann HC, Dunster J, Ko PK, Hu C (1996) MOSFET carrier mobility model based on gate oxide thickness, threshold and gate voltages. Solid-State Electronics 39:1515–1518
6. Jeon DS, Burk DE (1989) MOSFET electron inversion layer mobilities–a physically based semi-empirical model for a wide temperature range. IEEE Trans Electron Devices 36:1456–1463
7. Grabinski W, Bucher M, Sallese JM, Krummenacher F (2000) Compact modeling of ultra deep submicron CMOS devices. Int Conf on Signals and Electronic Systems 13–27
8. Fang FF, Fowler AB (1970) Hot electron effects and saturation velocities in Silicon inversion layers. J Appl Phys 41:1825–1831
9. Cheng Y et al (1997) Modelling temperature effects of quarter micrometer MOSFETs in BSIM3v3 for circuit simulation. Semicond Sci Technol 12:1349–1354
10. Pierret RF (1988) Semiconductor fundamentals, 2nd ed. Addison-Wesley, MA
11. Filanovsky IM, Allam A (2001) Mutual compensation of mobility and threshold voltage temperature effects with applications in CMOS circuits. IEEE Trans Circuits and Syst I: Fundamental Theory and Applications 48:876–884
12. Oxner ES (1988) FET technology and application. CRC Press, NY
13. Agarwal A, Mukhopadhyay S, Raychowdhury A, Roy K, Kim CH (2006) Leakage power analysis and reduction for nanoscale circuits. IEEE Micro 26:68–80
14. Fallah F, Pedram M (2005) Standby and active leakage current control and minimization in CMOS VLSI systems. IEICE Trans Electronics E88-C:509–519
15. Black JR (1969) Electromigration–a brief survey and some recent results. IEEE Trans Electron Devices 16:338–347
16. Ku JC, Ismail Y (2007) On the scaling of temperature-dependent effects. IEEE Trans Computer-Aided Design of Integrated Circuits and Syst 26:1882–1888
17. Morshed TH et al (2009) BSIM4.6.4 MOSFET model user's manual. [Online] http://www-device.eecs.berkeley.edu/~bsim3/bsim4_arch_ftp.html
18. Park C et al (1995) Reversal of temperature dependence of integrated circuits operating at very low voltages. Int Electron Devices Mtg 71–74
19. Bellaouar A, Fridi A, Elmasry MI, Itoh K (1998) Supply voltage scaling for temperature-insensitive CMOS circuit operation. IEEE Trans Circuits and Syst II: Analog and Digital Signal Processing 45:415–417
20. US Dept of Defense (2007) Integrated circuits (microcircuits) manufacturing, general specification, Std MIL-PRF-38535H. Washington DC
21. Sakurai T, Newton AR (1990) Alpha-power law MOSFET model and its applications to CMOS inverter delay and other formulas. IEEE J Solid-State Circuits 25:584–594
22. Shichman H, Hodges DA (1968) Modeling and simulation of insulated-gate field-effect transistor switching circuits. IEEE J Solid-State Circuits SC-3:285–289
23. Zhao W, Cao Y (2006) New generation of predictive technology model for sub-45nm early design exploration. IEEE Trans Electron Devices 53:2816–2823
24. Lasbouygues B, Wilson R, Azemard N, Maurine P (2006) Timing analysis in presence of supply voltage and temperature variations. ACM Int Symp on Physical Design 10–16
25. Kumar R, Kursun V (2006) Reversed temperature-dependent propagation delay characteristics in nanometer CMOS circuits. IEEE Trans Circuits and Syst II: Express Briefs 53:1078–1082

26. Dasdan A, Hom I (2006) Handling inverted temperature dependence in static timing analysis. ACM Trans Design Automation of Electronic Syst 11:306–324
27. Wolpert D, Ampadu P (2008) Normal and reverse temperature dependence in variation-tolerant nanoscale systems with high-k dielectrics and metal gates. 3rd ACM Int Conf on Nano-networks 1–5
28. Wong H, Iwai H (2006) On the scaling issues and high-k replacement of ultrathin gate dielectrics for nanoscale MOS transistors. Microelectronic Engineering 83:1867–1904
29. Langen D, Ruckert U (2002) Extending scaling theory by adequately considering velocity saturation. 15th Ann IEEE Int ASIC/SoC Conf 145–149
30. Zhao W (2008) Personal communication
31. Guillaumot B et al (2002) 75nm damascene metal gate and high-k integration for advanced CMOS devices. Int Electron Devices Mtg 355–358
32. Cheng B et al (1999) The impact of high-k gate dielectrics and metal gate electrodes on sub-100 nm MOSFETs. IEEE Trans Electron Devices 46:1537–1544
33. Sato T, Ichimiya J, Ono N, Hachiya K, Hashimoto M (2005) On-chip thermal gradient analysis and temperature flattening for SoC design. IEICE Trans Fundamentals E88-A:3382–3389
34. Tschanz J et al (2007) Adaptive frequency and biasing techniques for tolerance to dynamic temperature-voltage variations and aging. IEEE Int Solid-State Circuits Conf 292–604
35. Elgebaly M, Sachdev M (2007) Variation-aware adaptive voltage scaling system. IEEE Trans Very Large Scale Integr Syst 15:560–571
36. Martin S, Flautner K, Mudge T, Blaauw D (2002) Combined dynamic voltage scaling and adaptive body biasing for lower power microprocessors under dynamic workloads. IEEE/ACM Int Conf on Computer-Aided Design 721–725
37. Lee CC, de Groot J (2006) On the thermal stability margins of high-leakage current packaged devices. 8th Electronics Packaging Technology Conf 487–491
38. Kim NS et al (2003) Leakage current: Moore's law meets static power. Computer 36:68–75

Chapter 3
Sensing Temperature Dependence

In this chapter, we discuss the problem of temperature variation at the system level, and describe methods of detecting temperature and temperature dependence. In Sect. 3.1 we propose a low overhead adaptive temperature sensor, capable of adjusting its resolution and sampling rate based on the current temperature to reduce the energy required to maintain a chip thermal map. In addition, the proposed sensor features a process compensation unit, enabling the oscillator length to be adjusted to calibrate fabricated chips. In Sect. 3.2 we propose a temperature dependence sensor, capable of determining whether a core's I-T slope (temperature dependence) is positive or negative for a given supply voltage. The proposed sensor design was fabricated, and chip design issues and measurements are provided in Sect. 3.3.

3.1 Low Overhead, Energy-Efficient Adaptive Temperature Sensor

Variations in ambient and on-chip temperatures are increasingly common causes of delay errors in VLSI circuits. High ambient temperatures can cause path latencies to exceed clock periods, and intra-die temperature variations can result in communication errors between units with a large temperature differential.

To ensure reliability, multiple temperature sensors can create an on-chip heat map of both the chip temperature and the temperature differential between communicating units. An adaptive system can take this heat map information and adjust system parameters as necessary to guarantee functionality. In the Power5 microprocessor [1], 24 temperature sensors were used to generate this heat map. This large number of temperature sensors makes energy and area overhead of critical importance; unfortunately, many prior sensor designs focus on improving sensor resolution [2,3], while area and energy are lesser concerns. In this section, we present a low overhead, adaptive temperature sensor. Two adaptive techniques are explored to improve energy efficiency and enable runtime calibration to compensate for variations.

D. Wolpert and P. Ampadu, *Managing Temperature Effects in Nanoscale Adaptive Systems*, DOI 10.1007/978-1-4614-0748-5_3,
© Springer Science+Business Media, LLC 2012

3.1.1 Temperature Sensor Design

To tolerate temperature variations, a system can be designed to react to one or more temperature thresholds, $T_{threshold}$. A temperature sensor can provide the current chip temperature, T_{chip}, which will be used by the system to determine when these thresholds have been exceeded. In (3.1)–(3.3), $T_{guardband}$ is the temperature guardband which must be added to T_{chip} to ensure that the temperature does not pass $T_{threshold}$ in between readings.

$$T_{guardband} = \left(\frac{\partial T_{measured}}{\partial T} \right) \cdot t_{response} \qquad (3.1)$$

$$\partial T_{measured} = \partial T_{chip} + T_{accuracy} \qquad (3.2)$$

$$t_{response} = t_{sensor} + t_{correction} \qquad (3.3)$$

$\partial T_{measured} / \partial t$ is the measured temperature rate of change; $t_{response}$ is the system response time; $T_{accuracy}$ is the maximum change in temperature which might be hidden by sensor accuracy limitations; t_{sensor} is the sensor polling time; and $t_{correction}$ is the adaptation latency of the system once a change is detected. $T_{accuracy}$ and t_{sensor} determine how much energy the sensor consumes. $T_{accuracy}$ includes the sum of the maximum quantization noise (i.e. the sensor resolution) and any potential error which could be introduced by variations other than temperature.

Sensor resolution varies slightly depending on the chip temperature. The reported sensor resolutions in this work are the worst case resolution achieved over the entire −55°C to 125°C temperature range. Nonlinearities in the temperature sensitive element are included in this worst case resolution.

All simulations presented in this subsection use a 90 nm technology [4] with a supply voltage of 1 V. The proposed temperature sensor, shown in Fig. 3.1, consists of three main components: a ring oscillator, a pulse generator, and a pulse counter. The ring oscillator is the temperature sensitive component, and has been used extensively in prior art by measuring the temperature dependence of its operating frequency [5–7]. In the proposed work, the oscillator is gated with an enable signal, with some additional circuitry which will be explained momentarily. The pulse generator in Fig. 3.1 creates a full pulse for every transition of the oscillator, and can be used either to halve the sensor resolution for a small power increase, or nearly halve the overall power dissipation at a fixed resolution, as described in the next section.

The flip-flop chosen, used in the PowerPC 603 [8], is the most power and energy efficient option in a survey of flip-flop circuits [9]. The inverted output $\overline{Q}$ is created by adding an additional inverter after Q, shown in Fig. 3.2, rather than connecting $\overline{Q}$ directly to the output of the second transmission gate.

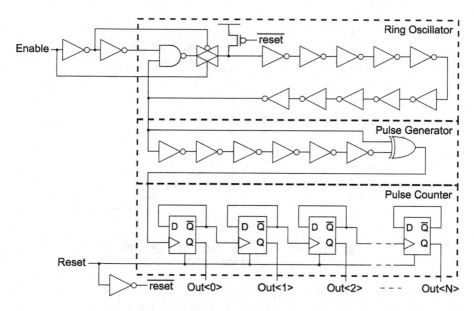

Fig. 3.1 Temperature sensor schematic

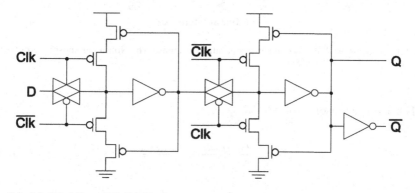

Fig. 3.2 Modified PowerPC 603 flip-flop

The number of flip flops, N_{FF}, required to support the $-55°C$–$125°C$ operating range is

$$N_{FF} = \lceil \log_2 (N_{pulse,\max}) \rceil \qquad (3.4)$$

$$N_{pulse,\max} = 2 \cdot f_{osc,\max} \cdot t_{enable} \qquad (3.5)$$

$N_{pulse,\max}$ is the maximum number of pulses generated at the output of the pulse generator, $f_{osc,\max}$ is the maximum oscillator frequency over the supported temperature range, and t_{enable} is the enable pulse width.

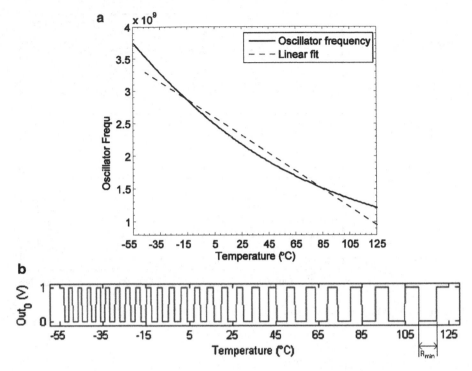

Fig. 3.3 (a) Ring oscillator nonlinear temperature response. (b) Nonlinear sensor output with R_{min} labeled

The average sensor resolution, R_{avg}, is

$$R_{avg} = \frac{T_{range,\max} - T_{range,\min}}{N_{pulse,\max} - N_{pulse,\min}} \tag{3.6}$$

where $N_{pulse,\min}$ is the minimum number of pulses generated by the pulse generator; $T_{range,\max}$ and $T_{range,\min}$ are the maximum and minimum temperatures in the supported range, respectively (measured in Kelvin). The average resolution is different from the minimum resolution, R_{min}, because f_{osc} varies nonlinearly with temperature (f_{osc} is shown in Fig. 3.3a with a linear fit). R_{min} is the largest range of temperatures over which the sensor produces the same value of N_{pulse}; R_{min} is labeled in Fig. 3.3b.

In the initial ring oscillator design, the enable signal was directly connected to the NAND gate, with no transmission gate or pull-up device. This resulted in the irregular output shown in the top half of Fig. 3.4 because the falling edge of the enable signal was not synchronized with the oscillator (synchronization is non-trivial because the oscillator frequency changes with temperature).

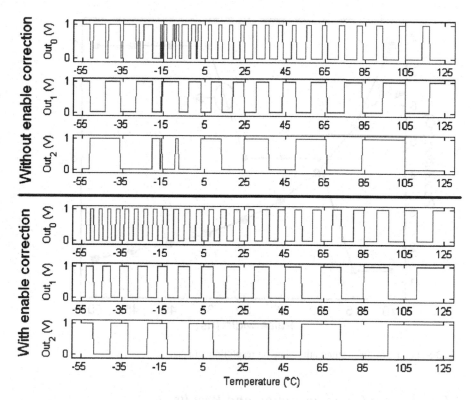

Fig. 3.4 Effect of the enable correction circuit on the sensor output

At certain temperatures, the lack of synchronization resulted in a rising edge just a few inverters behind a falling edge, causing the spurious behavior shown in the top half of Fig. 3.4 near −15°C. The locations and number of spurious pulses changes depending upon the enable pulse width. Furthermore, it results in an uneven duty cycle, decreasing the resolution.

The synchronization problem was eliminated by using a transmission gate to disconnect the NAND gate output before the spurious transition can occur. The resulting circuit is shown in Fig. 3.1, with the corrected behavior shown in the bottom half of Fig. 3.4. The reset device is necessary to ensure that the oscillator is in the correct state when re-enabled for the next sample, and also reduces any potential short circuit power which could result from the floating output node.

The output response of the sensor is nonlinear with temperature, causing the resolution to decrease as temperature increases (shown by the widening pulse widths in the lower half of Fig. 3.4). This nonlinearity is not problematic for our design, as long as the digital output associated with each temperature is consistent.

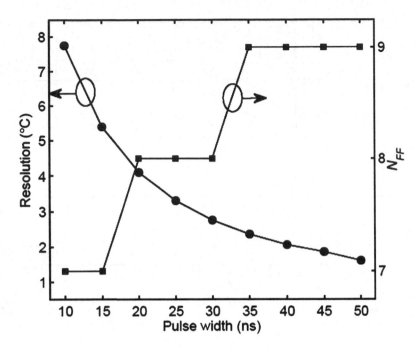

Fig. 3.5 Impact of the accumulation time on resolution and N_{FF}

3.1.2 Sensor Characterization and Results

The use of the enable signal provides an important capability—by extending the enable pulse width over multiple clock cycles, the temperature-induced delay change in the oscillator can be accumulated. This accumulation improves resolution, increasing the change in N_{pulse} between temperatures; unfortunately, the improvement comes at the expense of additional power dissipation as a result of the longer run time for each sample. A range of enable pulse widths between 10 and 50 ns is shown in Fig. 3.5 with the corresponding values of R_{min} and the minimum N_{FF} for each pulse width, calculated from (3.4).

In Fig. 3.6, the total energy dissipation per sample is shown. This energy dissipation includes a 1 ns reset pulse, 1 ns of settling time between the reset and enable signal, the enable pulse (of the indicated pulse width), and an additional 1 ns before reading out the data. As shown, larger pulse widths result in larger energy dissipation. Also of note is the significant increase in energy per sample as temperature is reduced. This is caused by the increased frequency of the oscillator at those lower temperatures.

The overall latency of the proposed sensor is PW + 3 ns, where PW is the enable pulse width. Thus, pulse widths between 10 and 50 ns in Fig. 3.5 correspond to sensor latencies between 13 and 53 ns, which are much smaller than latencies reported in prior temperature sensors. Another recent all-digital design [6] has a maximum

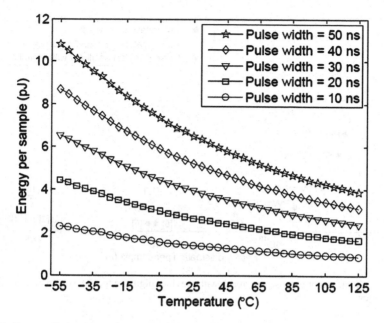

Fig. 3.6 Energy dissipation per sensor reading

sampling rate of 1 MHz, corresponding to a latency of 1 μs. The sensor in [6] achieved resolutions on the order of 0.5°C. Those results, achieved in a 0.35 μm technology, are scaled down to the 90 nm technology of the proposed design using a simple $1/S$ scaling rule [10], where S is the ratio between the two technology sizes. The scaled latency of the other work is 257 ns. For comparison, when a sample period of 257 ns (enable pulse width = 254 ns) is used in the proposed sensor, it also results in a resolution of ~0.5°C; however, the low overhead design of the proposed sensor results in an energy dissipation of just 37.6 pJ, while the energy dissipation scaled from [6] (energy scaling will be explained momentarily) is 8.57 nJ.

Figure 3.7 shows a scatterplot of the resolution and energy dissipation numbers reported by previous works, along with simulation results of a number of pulse width values in the proposed system. The energy-resolution trade-off in the proposed work can also be estimated mathematically by solving for t_{enable} using (3.5) and (3.6), and multiplying by the average power dissipation P_{avg} over the temperature range

$$E = t_{enable} \cdot P_{avg} = \frac{\left(T_{range,\max} - T_{range,\min}\right)}{2 \cdot R_{avg} \cdot \left(f_{osc,\max} - f_{osc,\min}\right)} \cdot P_{avg} \qquad (3.7)$$

Three types of systems are compared in Fig. 3.7: all-digital sensor implementations, making use of delay lines or ring oscillators with digital frequency conversion; analog sensor implementations, using current references with ΣΔ analog-to-digital conversion; and a mixed-signal implementation,

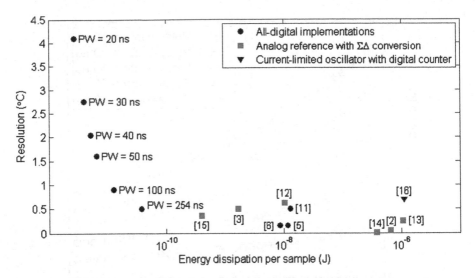

Fig. 3.7 Comparison of proposed approach with previous digital output sensors in terms of energy and accuracy

using a temperature-sensitive current bias to limit the speed of an oscillator with digital frequency conversion. The energy values of the previous works were found by inverting the sample rate to determine the sample time, and multiplying that time by the reported power dissipation. To improve the fairness of the comparison, the energy numbers from the previous works were all scaled down to their 90 nm equivalent using a $1/(U^2 S)$ rule, where U is the ratio between the supply voltages of the previous work and the proposed system supply voltage of 1 V. $1/U^2$ represents the scaling of the power dissipation, and $1/S$ represents the scaling of the latency [10]. Thus, the overall energy calculation for the previous works is

$$E = \frac{t_{sensor} \cdot P}{U^2 S} \tag{3.8}$$

Scaling likely underestimates the energy dissipations of the analog implementations, as the use of 90 nm for analog circuits introduces a number of reliability and accuracy difficulties. The technology node reported for each reference in Fig. 3.7 is listed in Table 3.1 along with the scaling factors.

The proposed design results in Fig. 3.7 show that sub-degree sensor accuracies can be achieved while also achieving significant energy savings. For a resolution of 4°C, the proposed method reduces energy dissipation by over two orders of magnitude compared to previous designs. Again, this large energy discrepancy with prior work is largely a result of the reduced resolution requirements and low overhead of the sensor design. The area overhead of the proposed design is 96 NAND2 equivalent gates for $N_{FF} = 8$. The fabricated sensor area will be discussed along with other fabrication details in Sect. 3.3.

Reference	Technology	$1/(U^2S)$
[2]	0.5 μm	0.014
[3]	0.6 μm	0.017
[5]	0.35 μm	0.024
[6]	0.35 μm	0.029
[11]	0.35 μm	0.029
[12]	2 μm	0.009
[13]	0.7 μm	0.014
[14]	0.7 μm	0.021
[15]	0.6 μm	0.014
[16]	80 nm	1.125

Table 3.1 Technology listing of scaled references

3.1.3 Adapting Sampling Rate and Resolution

Although the proposed design has already been shown to have excellent energy properties, we propose two additional techniques which take further advantage of the energy-resolution trade-off. These techniques can dramatically improve sensor energy dissipation depending upon the temperature conditions and desired temperature thresholds.

In the proposed sensor, total energy dissipation over a time window t, can be represented as

$$E(t) = t_{enable} \cdot r_{sample} \cdot P_{sample} + (1 - t_{enable} \cdot r_{sample}) \cdot P_{idle} \cdot t \qquad (3.9)$$

where r_{sample} is the sample rate, P_{sample} is the average power dissipated during the sampling process, and P_{idle} is the power dissipated when the sensor is idle (if the design is power gated, $P_{idle} \approx 0$ W).

Most sensors have a fixed sampling rate, using a relatively constant amount of energy (aside from the dependence on temperature). This can result in a large amount of wasted energy, because the temperature change over time is limited. To avoid wasting energy, r_{sample} can be adjusted depending upon the difference between the current temperature and the closest temperature threshold.

Temperature changes over time are limited by (1) the worst case power dissipation of the devices, (2) the thermal conductivity of the material, and (3) the thermal diffusivity of the material, according to [17]

$$dT(x, y, z, t) = \frac{\alpha_{th}}{k_{th}} \int_{t'=0}^{t} \frac{dP(t')}{(4\pi\alpha_{th}(t - t'))^{3/2}} \cdot e^{-\frac{r^2}{4\alpha_{th}(t-t')}} dt' \qquad (3.10)$$

$$dP(t') = q_V(t')dx'dy'dz' \qquad (3.11)$$

k_{th} is the thermal conductivity of the substrate (W/cm°C); $\alpha_{th} = k_{th}/\rho c$ is the thermal diffusivity (cm^2/s), where ρ is substrate density (g/cm^3) and c is specific heat (J/g°C);

t' is the time step; $dP(t')$ is the power dissipated at time t'; $q_V(t')$ is the power density per unit volume (W/cm^3); dx', dy', and dz' are the dimension of an elementary heat source; and r is the distance between the heat source and the point being measured.

Assuming that the distance between the temperature sensor and the heat source is negligible (i.e. $r = 0$), and that the power density over the time interval is constant (i.e. $q_v(t') = q_v$), (3.10) reduces to

$$dT(x, y, z, t) = \frac{\alpha_{th}}{k_{th}} \cdot \frac{P}{(4\pi\alpha_{th})^{3/2}} \cdot \frac{2}{\sqrt{t - t'}} \qquad (3.12)$$

Thus, with a silicon substrate ($k_{th} = 1.5 \cdot 10^{-4}$ W/μm°C, $\alpha_{th} = 9.3 \cdot 10^{-7}$ μm/s), assuming a power density of 90 W/cm^2 [18] over a 10×10 μm heat source, the maximum change in temperature over a 1 μs period is 0.28°C. Note that this is oversimplified for the purposes of example; the thermal impedance (°C/W) is also a function of time, causing the actual rate of temperature change to vary based on the time window [17].

Using this information, we can quantify the benefits of the adaptive polling rate. For example, consider a system where $T_{threshold} = 50°C$, and the maximum rate of temperature change is 0.28°C/μs. If $T_{chip} = 45°C$, then measurements must be taken at least every 17.86 μs to account for the worst case energy dissipation, which could increase the temperature by 5°C in that timeframe. If T_{chip} is reduced to $-20°C$, the temperature sensor polling rate can be reduced to ~250 μs/sample. This is a 93% reduction in r_{sample} compared to a fixed sampling rate of 17.86 μs/sample, and results in a 93% reduction in energy if we assume $P_{idle} = 0$.

This adaptive sensor polling technique is particularly useful with the proposed sensor because samples taken at low temperatures (where r_{sample} would be reduced) dissipate much more energy than samples taken at higher temperatures, as shown in Fig. 3.6. The energy dissipation of the adaptive approach compared to a fixed polling approach is shown in Fig. 3.8 for four different threshold temperatures. The overhead requirement for this technique is a look-up table to store the polling rates.

The proposed sensor accumulates error over a period of time, t_{enable}, which can also be adapted based on the temperature condition. As shown in (3.9), t_{enable} can be varied to trade off sensor resolution for additional energy savings. There are two additional overhead requirements for this method: an adjustable enable circuit for varying the number of clock cycles over which the error is accumulated, and an additional circuit to compare the output vector of each t_{enable} to the corresponding $T_{threshold}$ vector (the output vectors change with the enable pulse width, thus a $T_{threshold}$ vector must be stored for each potential pulse width).

To use the previous example, assume $T_{threshold} = 50°C$. Assume that two pulse widths are available, 20 and 10 ns, corresponding to a 4°C resolution and an 8°C resolution, respectively. In this example, the polling rate is fixed such that temperature can change by no more than 4°C between samples. If $T_{chip} \geq 38°C$

Fig. 3.8 Energy savings of adaptive sensor polling

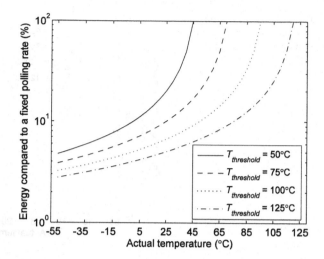

(=50°C threshold – 8°C resolution – 4°C max. temperature change), t_{enable} must be set to 20 ns; however, when $T_{chip} < 38°C$, t_{enable} can be set to 10 ns, reducing sensor energy by ~50%.

Using the earlier example where $T_{threshold} = 50°C$ and $T_{chip} = -20°C$, we can calculate the effective energy savings when the two adaptive approaches are combined. When $T_{chip} < 38°C$, the resolution is set to 8°C. With this resolution, the adaptive sampling rate must be large enough to detect a 58°C (=$38°C - T_{chip}$) change in temperature, corresponding to $r_{sample} = 207$ μs. Compared to a fixed polling scheme with $r_{sample} = 17.86$ μs and a fixed resolution of 4°C, the combination of the two adaptive schemes results in energy savings of 52% compared to the adaptive sampling rate alone, for a total energy savings of 96.5% compared to a fixed sensor design. The range of potential energy savings based on the combination of these two adaptive approaches versus a fixed approach is shown in Fig. 3.9 for four different temperature thresholds, with a maximum energy savings of 98.6% occurring when the actual temperature is $-55°C$ and $T_{threshold}$ is 125°C.

3.1.4 Process Compensation Unit

The enable pulse width of the temperature sensor may be varied after fabrication as a first order accuracy compensation. The further away the actual temperature is from the compensation point (commonly set to the middle of the temperature range), the more susceptible the output vector is to process and voltage variations. Supply voltage-induced variations in oscillator frequency can be effectively addressed through the use of decoupling capacitors or a reduced supply bounce circuit, which have been used to reduce supply voltage variations below 2% [19]. Unfortunately, available techniques for controlling process variation are considerably less effective, with V_T variations of $\pm 10\%$ still common [20]. To compensate for process variations, we propose the process compensation unit shown in Fig. 3.10.

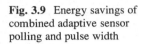

Fig. 3.9 Energy savings of combined adaptive sensor polling and pulse width

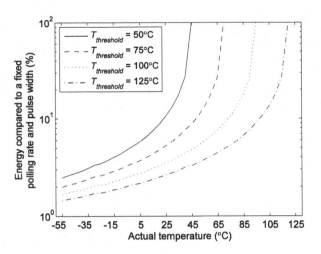

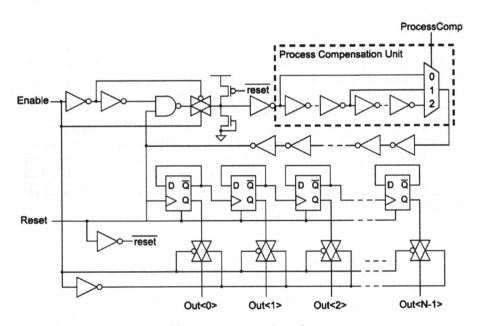

Fig. 3.10 Temperature sensor with process compensation unit

Variations in V_T can be determined during post-fabrication testing, and this information can be used to compensate for the process-induced changes in f_{osc}. We exploit this process information as an input to a process compensation unit (the *ProcessComp* signal in Fig. 3.10) that adjusts the number of stages in the oscillator to match a target f_{osc}.

3.2 A Temperature Dependence Sensor

Before describing how to sense the temperature dependence, let us briefly summarize the temperature dependence description from Chaps. 1 and 2. Changes in temperature affect system speed, power, and reliability by altering the threshold voltage V_T and mobility μ in each device [21]. The resulting changes in device current can lead to timing failure or cause circuits to exceed power or energy budgets. The impact of temperature on device current depends on the supply voltage. Near a technology's nominal voltage, the current-temperature (*I-T*) dependence is negative (also referred to as the normal temperature dependence)—drain current I_D (and device speed) *decreases* with increasing temperature. At lower voltages, the current-temperature dependence is positive (the reverse dependence), and I_D *increases* with increasing temperature [22].

In circuits with negative *I-T* dependences, timing failures will occur at high temperatures; however, in circuits with positive *I-T* dependences, timing failures will occur at *low* temperatures. Positive *I-T* dependences approach nominal voltages as technology scales [23,24]; thus, adaptive systems that vary supply voltage to reduce energy consumption or improve reliability [25,26] will have operating voltages in both the negative and positive *I-T* regions.

Existing temperature sensors assume that a circuit's *I-T* dependence is negative, which will cause three problems—(1) overheating circuits and timing failures may go undetected in the positive *I-T* dependence region, (2) sensors may generate false positives and unnecessarily reduce performance in the positive *I-T* dependence region, and (3) excessive frequency guardbands will be required to ensure reliable operation across both *I-T* dependences, wasting throughput or energy. The type of problem will depend on the sensor design—some temperature sensors track circuit delay, while others measure device voltages or currents to determine the temperature. If the temperature dependence is not known, sensors that do not track circuit delay may misdiagnose timing failures, while sensors that do track circuit delay may misdiagnose overheating circuits; for example, a *decrease* in delay could mean (1) a circuit is in the negative *I-T* region and *is not* overheating, or (2) the circuit is in the positive *I-T* region *is* overheating. Thus, including the positive *I-T* dependence in sensor design is critical for detecting overheating circuits and timing failures.

To solve these problems, this section presents a sensor which determines whether a system is in the normal or reverse temperature dependence region.

3.2.1 Temperature Dependence Sensor Design

To detect timing failures and overheating at negative *I-T* slopes, temperature sensors uses a look-up table with pre-determined thresholds [25,27]; when the temperature *exceeds* a threshold, cooling systems can be triggered or frequency can be throttled. For positive *I-T* slopes, the proposed temperature dependence sensor system output can also trigger the performance throttle when delay *reduces below* a threshold.

To use different trigger points for the negative and positive *I-T* dependence regions, we must be able to determine the dependence region of the circuit. Determining the temperature dependence of a circuit is more complex than determining a static temperature. The dependence *TempDep* can be represented as

$$TempDep = \begin{cases} 1, & \frac{\partial \tau}{\partial T} \geq 0 \\ 0, & \frac{\partial \tau}{\partial T} < 0 \end{cases} \tag{3.13}$$

where $\partial \tau / \partial T$ is the change in delay corresponding to a change in temperature.

Calculation of the temperature dependence requires a change in temperature to make a reading. If the sensor is in an environment where temperature changes very slowly, the sample time may have to be reduced before the dependence sensor is able to provide a reading. For system design, this is of limited importance; if the temperature is not changing, there is no temperature impact on delay.

The temperature dependence can be determined using two component temperature sensors; one at the same operating voltage as the circuit being monitored, V_{OP}, and a temperature reference at a voltage V_{REF} that will maintain the sign of its *I-T* slope despite process or voltage variations. V_{REF} is a static voltage, and may be generated externally or by an internal DC-DC converter. If the two sensor readings change in the same direction, the dependence at V_{OP} matches the known dependence at V_{REF}; if the readings change in opposite directions, the dependence is the opposite of the known dependence. Thus, the dependence sensor output *Sens_Out* can be represented as follows:

$$Sens\_Out = \begin{cases} 1, & (TempDep)_{V_{DD}} \neq (TempDep)_{V_{REF}} \\ 0, & (TempDep)_{V_{DD}} = (TempDep)_{V_{REF}} \end{cases} \tag{3.14}$$

The temperature dependence sensor implementation is shown in Fig. 3.11. Two component temperature sensors are used which simultaneously take readings at the chosen V_{REF} and the operating voltage of the adaptive system, V_{OP} (these readings are referred to as $T_{REF,t}$ and $T_{OP,t}$, respectively). Low-voltage up shifters [28], shown in Fig. 3.12, are used to convert the lower voltage component sensor output to the higher component sensor voltage to facilitate comparison.

Each reading is compared with its previous sample, $T_{REF,t-1}$ and $T_{OP,t-1}$, respectively, which are stored in buffers. The comparators consist of the overflow circuit of a kill-zero carry lookahead adder (shown in Fig. 3.13) and indicate if $T_t \geq T_{t-1}$ or $T_t < T_{t-1}$. The two comparator outputs are then fed into an XOR gate, which determines if the temperature dependence at V_{OP} is the known dependence or if it is the opposite dependence.

As mentioned, if the temperature does not change (or the change is too small to affect the temperature sensor outputs), the comparator outputs will indicate

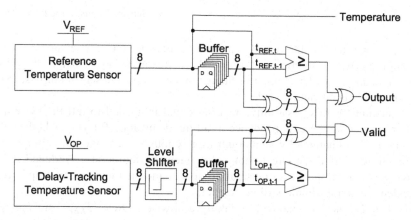

Fig. 3.11 Temperature dependence sensor schematic

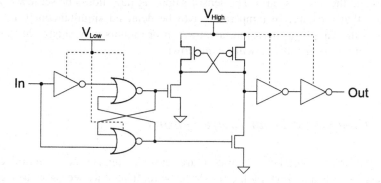

Fig. 3.12 Circuit design of low-voltage level shifter

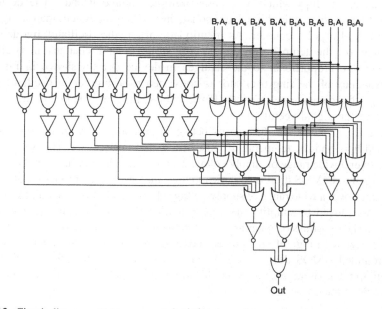

Fig. 3.13 Circuit diagram of kill-zero carry lookahead overflow comparator

$T_t \geq T_{t-1}$ regardless of the actual temperature dependence. For example, if the temperature sensor resolution is $1°C$ and the temperature only changes by $0.5°C$ between readings, there may not be a change in the component sensor outputs, so no dependence decision can be made.

To account for this possibility, a *Valid* signal is needed to tell the system if the current temperature dependence reading is meaningful (i.e. whether the temperature has changed by an amount large enough to change both component sensor outputs). If a reading is not valid, then the temperature-induced delay change between samples was too small to detect, and the most recent temperature dependence reading should still be correct.

The *Valid* signal is generated by taking a bitwise XOR of $T_{REF,t}$ and $T_{REF,t-1}$, and a bitwise XOR of $T_{OP,t}$ and $T_{OP,t-1}$. These vectors are reduced to one bit outputs with an OR tree, and passed into an AND gate to check if both sensor outputs have changed and the result is valid. The sensor sampling rate should be set low enough to ensure that a change in temperature can be detected simultaneously by both sensors; if the sampling rate is too fast, the sensor outputs may change in separate readings, and the dependence will be unknown.

3.2.2   Component Temperature Sensors

Our low-complexity component temperature sensor design is used to minimize area and energy overhead, allowing the sensor to be replicated as necessary to create a comprehensive thermal map of a chip, as well as determining multiple temperature dependences in chips which may have multiple voltage islands. The design is similar to that shown in Fig. 3.10, consisting of a process-compensating ring oscillator and a pulse counter. As mentioned earlier, the oscillator frequency is dependent on the temperature and supply voltage, and changes in oscillator frequency are proportional to the changes in device current shown in Fig. 2.6. The frequency of the ring oscillator is converted to a number of oscillations by applying a fixed enable pulse width (PW), and this number is stored in the counter to produce the digital vector *Out*.

There are a few minor differences between the component temperature sensor presented here and the temperature sensor presented in Sect. 3.1. This sensor must operate over a wide range of supply voltages, and so the flip flops used in the counter are changed to C^2MOS flip-flops, shown to have the best energy properties of a range of flip-flops over a wide range of supply voltages [29]. In addition, the pulse generator has been removed to simplify the sensor characterization over the wide range of voltages. The transmission gates are added to the outputs of the pulse counter to avoid unnecessary toggling of the other components in the dependence sensor. The grounded NMOS device ensures that leakage through the *Reset* device does not pull up the floating node between the end of the enable pulse and the time when the sample is read.

Table 3.2 Characterization of the temperature dependence sensor

Unit	Area (NAND2 equiv. gates)	Energy per sample (pJ)
Temperature sensor (x2)	212	7.3
8-bit level shifter	10	0.14
8-bit buffer (x2)	159	0.17
Comparator (x2)	89	0.31
8-bit bitwise XOR (x2)	26	0.09
Output XOR and NAND	3	0.01
Total	985	15.89

PW can be varied after fabrication as a first order accuracy compensation. The further away the actual temperature is from the compensation point, the more susceptible the output vector is to process variation (fluctuations in supply voltage should be minimized with decoupling capacitors, which may be non-trivial in these multi-voltage systems [30]). For example, if the system is expected to fail at 85°C, the desired *Out* vector can be tuned to that temperature to optimize accuracy.

3.2.3 Dependence Sensor Characterization

The dependence sensor was simulated in a 90 nm technology. Area and energy dissipation numbers of each component are provided in Table 3.2 for a supply voltage of 1 V and PW = 50 ns. As shown, the component temperature sensors consume about 92% of the total energy per sample. As mentioned in the last subsection, the transmission gates in the component sensors disconnect the *Out* vector from the rest of the dependence sensor when the oscillator is enabled, eliminating unnecessary toggling of the other units.

The component sensor resolution and energy dissipation per sample are a function of PW and supply voltage, shown in Fig. 3.14 for a fixed supply voltage of 1 V, and in Fig. 3.15 for varying supply voltages with a fixed temperature resolution of 1°C. In Fig. 3.14, an increase in PW results in a proportional increase in energy. The majority of the energy dissipation occurs when the ring oscillator is active, resulting in a linear relationship between energy and PW. Temperature variations occur on the order of 10^{-4} s [31], thus the sensor sampling frequency can be in the kHz range. The sensor can be power-gated in between readings to avoid wasting energy from leakage.

Figure 3.15 shows the required PW to guarantee a 1°C resolution across a range of supply voltages, and the corresponding energy for a 1 kHz sampling rate at each voltage. The nonlinear behavior in Fig. 3.15 is a result of two opposing factors—increasing PW increases energy dissipation, while decreasing supply voltage decreases energy dissipation. The supply voltage effect is shown to dominate the PW effect below voltages of ~0.8 V.

Figure 3.16a shows the simulated dependence sensor operation, generated by measuring the component sensor outputs at 30°C, inserting those values into

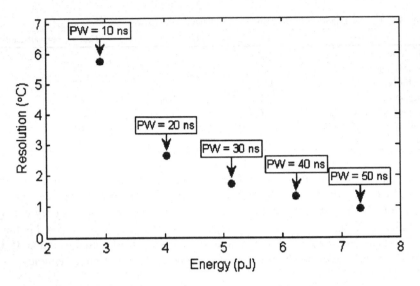

Fig. 3.14 Impact of the sensor enable pulse width on energy and resolution

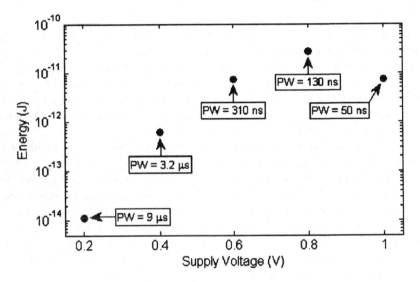

Fig. 3.15 Impact of the supply voltage on energy for a 1°C resolution

the buffers in Fig. 3.11, and then simulating the entire system at 70°C to collect the dependence sensor output. The dependence sensor output is equal to 1 V when the dependence is the same as the dependence at V_{REF}, and equal to 0 V when the dependence is the opposite of that at V_{REF}. V_{REF} is set to 1 V and V_{OP} is varied, indicating a switch from the normal dependence region to the reverse

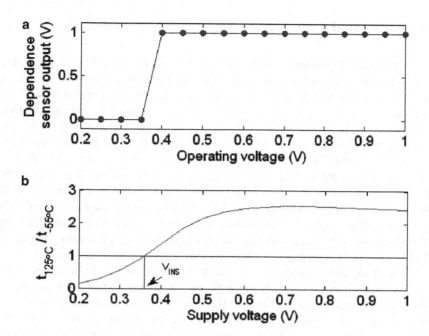

Fig. 3.16 (a) Operation of the dependence sensor. (b) Percentage change in inverter delay from 125°C to −55°C

dependence region between $V_{OP} = 0.4$ V and $V_{OP} = 0.45$ V. Figure 3.16b shows the change in delay of an inverter between 125°C and −55°C, with V_{INS} labeled where the delays are equal. Figure 3.16b indicates that the sensor is accurate to within 10% of V_{INS} ($V_{INS} \cong 360$ mV) in a 90 nm technology. Within this 10%, the change in delay caused by temperature over the −55–125°C temperature range is 14.8%.

Clearly, the presented temperature sensors are also susceptible to process and voltage variation, which will impact the energy and accuracy of the proposed design. For a pulse width of 50 ns, a ±10% change in V_{DD} changes a temperature reading by up to 12%, and a ±10% change in V_T changes a reading by up to 9%. An increase in V_{DD} (or decrease in V_T) in the −I-T slope region will make the temperature appear to decrease; in the +I-T slope region, the temperature will appear to increase. If process- or voltage-induced delay changes cause one component sensor output to change in the opposite direction of an actual change in temperature, the dependence sensor output will be erroneous; however, if both component sensors are affected by the change, the dependence sensor output will be correct. Thus, if the component sensors are placed in close proximity, the impact of process and voltage variation on the overall temperature dependence output will be reduced. The necessary guardbands to prevent process- or voltage-induced failure are examined in Chap. 4.

3.3 Fabricated Temperature Dependence Sensor Chip

The temperature dependence sensor implementation in Fig. 3.11 was fabricated in a 0.35 μm technology (the chip micrograph and layout are shown in Fig. 3.17 [32]).The temperature reference sensor can be implemented using any temperature sensor design, although the sensor at V_{OP} must be a delay-tracking temperature sensor to detect temperature-induced timing failures. For simplicity, this implementation uses the same sensor design for both the temperature reference and the sensor at V_{OP}. Consecutive readings at V_{REF} are compared to determine if the chip temperature is increasing or decreasing. Comparing the direction of the change at V_{REF} to the change in readings at V_{OP} (using an XOR gate) determines the I-T dependence at V_{OP}. Level converters are used to convert the lower voltage sensor's *Out* vector to the high sensor voltage to facilitate comparison. Both the V_{REF} and V_{OP} sensors have 8-bit outputs, achieving up to a 0.29°C resolution over the 5–80°C range of measured temperatures.

Temperature measurements were made using a custom-built temperature apparatus, shown in Fig. 3.18, using a Ni-Fe thermistor with ±0.1°C accuracy. 5 W 100 Ω resistors were used to heat the enclosed space up to 80°C (measured internally in 2°C increments by attaching the thermistor to the chip), and temperatures down to 5°C were measured using an ice bath (also measured internally). System performance is summarized in Table 3.3.

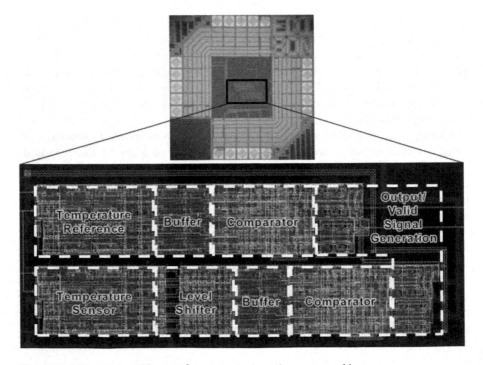

Fig. 3.17 Micrograph and layout of temperature dependence sensor chip

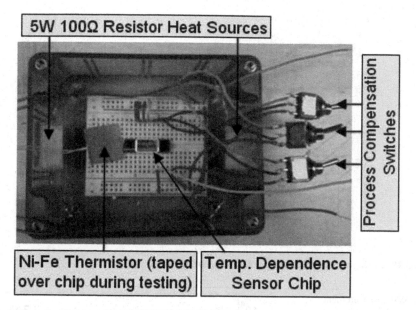

Fig. 3.18 Temperature testing apparatus

Table 3.3 Fabricated temperature dependence sensor performance summary

Parameter	Value
Technology	0.35 μm
Supply voltage	0.5–3.3 V
Area	0.049 mm^2
Temperature range	5–80°C
Resolution	0.29°C
Accuracy	±1.3°C
Sample latency	20 μs
Avg. sample power	15.5 mW
Energy per sample	310.0 nJ

3.3.1 Measurement of Sensor System Functionality

System functionality is shown over a wide voltage range in Fig. 3.19. In the 0.35 μm technology, the positive temperature dependence occurs at voltages below ~1.4 V, shown by simulation in Fig. 3.19a; thus, the measurements in Fig. 3.19b show that the sensor system correctly indicates the transition between the positive and negative *I-T* slopes. Further measurements of the impact of supply voltage on temperature variation are shown in Fig. 3.20, indicating the importance of considering the positive *I-T* slope at low voltages; at 0.5 V, temperature affects f_{osc} by more than an order of magnitude over the measured temperature range. Figure 3.21 indicates the change in temperature ΔT needed to ensure a valid reading at each supply voltage (recall that the enable period *En* impacts the resolution of the component sensors). Lower

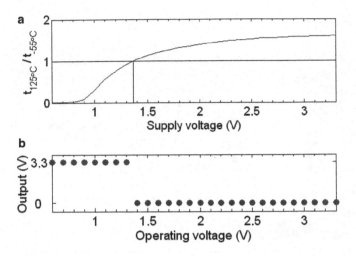

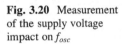

Fig. 3.19 (a) Simulated temperature sensitivity. (b) Measured operation

Fig. 3.20 Measurement of the supply voltage impact on f_{osc}

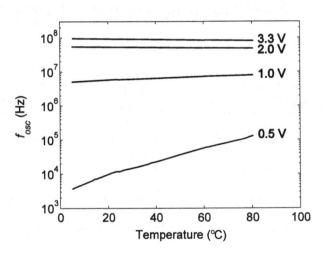

voltages require larger ΔTs or larger enable periods to make a valid reading. *En* may also be adjusted to trade-off resolution and energy consumption.

3.3.2   Measurement of Sensor System Accuracy and Process Compensation

Measurements of the linearity of the oscillator frequency f_{osc} are provided in Fig. 3.22 for six measured chips; linearity offsets as low as 0.2% (corresponding to an accuracy of 1.3°C, including both sensor inaccuracies and measurement inaccuracies) were achieved over the 5–80°C temperature range, although inter-die

Fig. 3.21 Measured ΔT needed for a valid reading

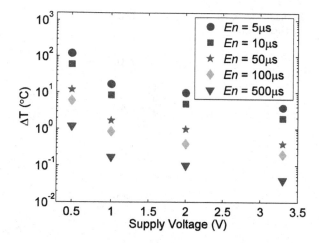

Fig. 3.22 Measured f_{osc} temperature linearity

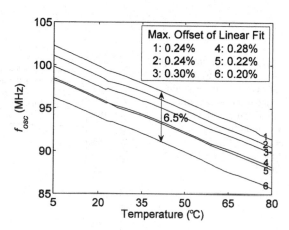

process variation resulted in f_{osc} shifts of up to 6.5%. The sensor system output uses the difference between readings of the temperature sensor and temperature reference; thus, if the temperature reference and sensor are placed in close proximity, the impact of process and voltage variations on the overall temperature dependence sensor output will be reduced.

However, if the reference or sensor is also used to find the temperature, they must be calibrated. The tuning range of the process compensation unit is shown in Fig. 3.23, indicating that the chip is capable of a larger compensation (over ±10%) than the measured variations. Post-fabrication tuning was achieved using the manual switches shown in Fig. 3.18 for simplicity. In this design, a 35-stage oscillator was used with a ±4 stage range selectable in 2-stage increments. A larger oscillator would enable a fine tuning range and enable greater calibration accuracy.

Fig. 3.23 Measured process compensation range

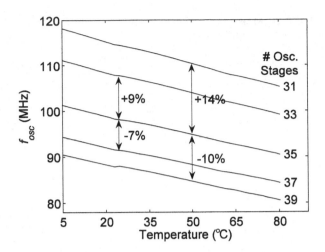

Fig. 3.24 Adjustable length ring oscillator used for post-fabrication compensation reduces the impact of a simulated ±10% process variation from +10.1%/−12.3% (*solid lines*) to +3.6%/−2.2% (*dashed lines*)

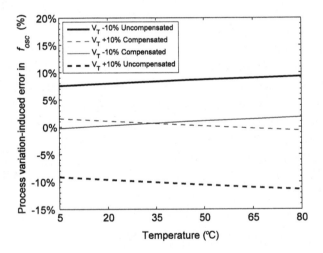

We also simulate the effectiveness of our compensation unit for a larger ±10% process variation range in Fig. 3.24. The solid curves in Fig. 3.24 are for simulations of ±10% V_T variations. The y-axis shows the impact of those variations on f_{osc} in a 35-stage oscillator, varying up to +10.1% and −12.3% for the −10% and +10% V_T cases, respectively. Reducing the number of stages in the oscillator increases f_{osc}; for the +10% V_T case, an increase of four stages reduces the process-induced error to just 3.6%. Similarly, increasing the number of stages in the oscillator compensates for reductions in V_T, reducing the process-induced error in the −10% V_T case to just 2.9%.

Table 3.4 Performance comparison with other temperature sensors

Design	Sensor type	Area (mm^2)	Energy per sample (μJ)	Tech. node	Accuracy (°C)	Detects timing failures at		Detects overheating at	
						+I-T	–I-T	+I-T	–I-T
[5]	Delay-tracking	0.175	0.5	0.35 μm	−0.7/+0.9	✓	✓		✓
[16]	Delay-tracking	0.02	1.0	80 nm	N/A	✓	✓		✓
[34]	Delay-tracking	0.05	0.003	0.18 μm	−1.6/+3	✓	✓		✓
[35]	Voltage-based	4.5	6.3	0.7 μm	±0.25		✓	✓	✓
[36]	Voltage-based	0.02	1.6	32 nm	±5		✓	✓	✓
[37]	Current-based	0.01	N/A	0.35 μm	±1.97		✓	✓	✓
Prop.	Delay-tracking	0.049	0.3	0.35 μm	±1.3	✓	✓	✓	✓

3.3.3 Comparison with Other Temperature Sensors

The proposed system is compared with other temperature sensors in Table 3.4. While delay-based sensors can detect timing failures for any I-T curve and voltage- or leakage-based sensors can detect overheating for any I-T curve, the proposed sensor system is able to detect both overheating *and* timing failures for any I-T curve. This improved functionality is achieved while maintaining area, accuracy and energy values similar to other sensor designs; using a linear technology scaling coefficient, the proposed system occupies the smaller area after the leakage current-based sensor, with energy dissipation only a small fraction of the voltage-based sensors. The additional functionality and competitive performance make the proposed sensor system a useful addition for reliable system design. Note that the compared sensors all examine device temperature; additional work has been proposed that examines the temperature of on-chip interconnect wires [33], enabling a more holistic view of a chip's temperature conditions.

3.4 Summary

Managing the impact of temperature on circuit functionality becomes increasingly complex as technology scales. To avoid overheating and timing failures in multi-voltage systems, awareness of the I-T dependence at each voltage (and process point) is critical. This paper has presented the first fabricated temperature dependence sensor to efficiently and accurately detect both positive and negative *I-T* slopes. This dependence sensing will be critical for handling temperature-related delay changes as the positive temperature effect becomes observable closer to nominal voltages. Furthermore, the proposed process compensation method can help maintain sensor accuracy and functionality despite process variations or device aging. The dependence sensor is accurate to within 1.3°C and dissipates 310 nJ per sample at a supply voltage of 3.3 V with a 20 µs sample latency.

References

1. Clabes J et al (2004) Design and implementation of the POWER5 microprocessor. IEEE Int Solid-State Circuits Conf 56–57
2. Pertijs M, Makinwa K, Huijsing JH (2005) A CMOS smart temperature sensor with a 3σ inaccuracy of ±0.1°C from −55°C to 125°C. IEEE J Solid-State Circuits 40:2805–2815
3. Tuthill M (1998) A switched-current, switched-capacitor temperature sensor in 0.6-um CMOS. IEEE J Solid-State Circuits 33:1117–1122
4. Zhao W, Cao Y (2006) New generation of predictive technology model for sub-45 nm early design exploration. IEEE Trans Electron Devices 53:2816–2823
5. Chen P, Chen CC, Tsai CC, Lu WF (2005) A time-to-digital-converter-based CMOS smart temperature sensor. IEEE J Solid-State Circuits 40:1642–1648
6. Chen CC, Chen P, Liu AW, Lu WF, Chang YC (2006) An accurate CMOS delay-line-based smart temperature sensor for low-power low-cost systems. Measurement Science and Technology 17:840–846
7. Quenot GM, Paris N, Zavidovique B (1991) A temperature and voltage measurement cell for VLSI circuits. Euro ASIC 334–338
8. Gerosa G et al (1994) A 2.2 W, 80 MHz superscalar RISC microprocessor. IEEE J Solid-State Circuits 29:1440–1454
9. Stojanovic V, Oklobszija V (1999) Comparative analysis of master-slave latches and flip-flops for high-performance and low-power systems. IEEE J Solid-State Circuits 34:536–548
10. Rabaey J, Chandrakasan A, Nikolic B (2003) Digital Integrated Circuits: A Design Perspective, 2nd ed. Prentice Hall, NJ
11. Chen P, Chen CC, Chen TK, Chen SW (2006) A time domain mixed-mode temperature sensor with digital set-point programming. IEEE Custom Integrated Circuits Conf 821–824
12. Bakker A, Huijsing JH (1996) Micropower CMOS temperature sensor with digital output. IEEE J Solid-State Circuits 31:933–937
13. Bakker A, Huijsing JH (1999) A low-cost high-accuracy CMOS smart temperature sensor. IEEE European Solid-State Circuits Conf 302–305
14. Pertijs MAP et al (2005) A CMOS smart temperature sensor with a 3σ inaccuracy of ±0.5°C from −50°C to 120°C. IEEE J Solid-State Circuits 40:454–461
15. Luh L Jr, Choma J, Draper J, Chiueh H (1999) A high-speed CMOS on-chip temperature sensor. IEEE European Solid-State Circuits Conf 290–293
16. Kim CK, Lee JG, Jun YH, Lee CG, Kong BS (2007) CMOS temperature sensor with ring oscillator for mobile DRAM self-refresh control. Microelectronics 38:1042–1049
17. Rinaldi N (2001) On the modeling of the transient thermal behavior of semiconductor devices. IEEE Trans Electron Devices 48:2796–2802
18. Rusu S (2001) Trends and challenges in VLSI technology scaling towards 100 nm. European Solid-State Circuits Conf 194–196

19. Ji G, Arabi T, Taylor G (2005) Design and validation of a power supply noise reduction technique. IEEE Trans on Adv Packaging 28:445–448
20. Rabaey J (2009) Low Power Design Essentials. Springer, USA
21. Filanovsky IM, Allam A (2001) Mutual compensation of mobility and threshold voltage temperature effects with applications in CMOS circuits. IEEE Trans Circuits and Syst I: Fundamental Theory and Applications 48:876–884
22. Park C et al (1995) Reversal of temperature dependence of integrated circuits operating at very low voltages. Int Electron Devices Mtg 71–74
23. Kumar R, Kursun V (2006) Reversed temperature-dependent propagation delay characteristics in nanometer CMOS circuits. IEEE Trans Circuits and Syst II: Express Briefs 53:1078–1082
24. Wolpert D, Ampadu P (2008) Normal and reverse temperature dependence in variation-tolerant nanoscale systems with high-k dielectrics and metal gates. 3rd ACM Int Conf on Nano-networks 1–5
25. Tschanz J et al (2007) Adaptive frequency and biasing techniques for tolerance to dynamic temperature-voltage variations and aging. IEEE Int Solid-State Circuits Conf 292–604
26. Elgebaly M, Sachdev M (2007) Variation-aware adaptive voltage scaling system. IEEE Trans. Very Large Scale Integr (VLSI) Syst 15:560–571
27. Wolpert D, Fu B, Ampadu P (2010) Temperature-aware delay borrowing for energy-efficient low-voltage link design. 4th ACM/IEEE Int Symp on Networks-on-Chip 107–114
28. Hass KJ, Cox DF (2000) Level shifting interfaces for low voltage logic. 9th NASA Symp on VLSI Design 3.1.1-3.1.7
29. Chavan A, Dukle G, Graniello B, MacDonald E (2006) Robust ultra-low power subthreshold logic flip-flop design for reconfigurable architectures. IEEE Int Conf on Reconfigurable Computing and FPGAs 1–7
30. Popovich M, Friedman EG (2006) Decoupling capacitors for multi-voltage power distribution systems. IEEE Trans Very Large Scale Integr (VLSI) Syst 14:217–228
31. Bernstein K et al (2006) High-performance CMOS variability in the 65-nm regime and beyond. IBM J Res and Dev 50:433–449
32. Wolpert D, Ampadu P (2011) A sensor system to detect positive and negative temperature dependences. IEEE Trans Circuits and Syst II: Express Briefs 58:235–239
33. Datta B, Burleson W (2008) Collaborative sensing of on-chip wire temperatures using interconnect-based ring oscillators. 18th ACM Great Lakes Symp on VLSI 41–46
34. Lin YS, Sylvester D, Blaauw D (2008) An ultra low power 1 V, 220nW temperature sensor for passive wireless applications. IEEE 2008 Custom Integrated Circuits Conf 2661–2668
35. Aita AL, Pertijs M, Makinwa K, Huijsing JH (2009) A CMOS smart temperature sensor with a batch-calibrated inaccuracy of $\pm 0.25°C$ (3σ) from $-70°C$ to $130°C$. IEEE Int Solid-State Circuits Conf 342–343
36. Lakdawala H, Li YW, Raychowdhury A, Taylor G, Soumyanath K (2009) A 1.05 V 1.6 mW, 0.45°C 3σ resolution $\Sigma\Delta$-based temperature sensor with parasitic resistance compensation in 32 nm digital CMOS process. IEEE J Solid-State Circ 44:3621–3630
37. Ituero P, Ayala JL, Lopez-Vallejo M (2008) A nanoWatt smart temperature sensor for dynamic thermal management. IEEE Sensors J 8:2036–2043

Chapter 4
Variation-Tolerant Adaptive Voltage Systems

Runtime variations in voltage and temperature can cause large changes in latency, resulting in functional failure if appropriate design margins are not maintained. With temperature fluctuations up to 50°C [1] and supply voltage variations of 22% [2], frequency guardbands can become quite large, limiting system throughput. Adaptive solutions to optimize guardbands can allow systems to tolerate extreme worst-case scenarios without sacrificing delay and power.

Power dissipation can be effectively addressed through a variety of techniques including clock gating [3], power gating [4], and adaptive voltage scaling (AVS) [5]. While clock and power gating involve shutting down sections of a chip to reduce switching and leakage power, respectively, AVS allows for circuits to continue functioning at a slower speed. By reducing supply voltage when lower throughput is required, AVS addresses both switching and leakage power. Unfortunately, the multiple supply voltages of an AVS system also exacerbate process, voltage, and temperature (PVT) variation problems.

Process variations are generally the result of lithography limitations and quantum effects in oxides and doping profiles [6]. Voltage variation is caused by distribution networks with inconsistent IR drops and local voltage bounces due to fluctuating input patterns [7]. Temperature variations have both on-chip and environmental sources. Across-chip variations can result in sub-blocks with different PVT parameters, which can result in large delay differences and cause communication failures [1]. Voltage variations are particularly important because many adaptive systems control applied voltages [8–11]; in these systems, any voltage instability could result in inaccurate or undesirable speed and power adjustments. To ensure system robustness while avoiding over-conservative design, runtime variations should be considered with respect to local PVT variations. The interaction of these types of variation is the focus of this chapter.

Two well-known adaptive techniques to compensate for variation are AVS and adaptive body bias (ABB) [8, 9]. These methods have been studied without considering variations, mathematically analyzing voltage and body bias points for optimal energy performance [9]. They have also been used to consider variation in a variety of ways, such as restoring performance when variation thresholds are

D. Wolpert and P. Ampadu, *Managing Temperature Effects in Nanoscale Adaptive Systems*, DOI 10.1007/978-1-4614-0748-5_4,
© Springer Science+Business Media, LLC 2012

crossed [8], improving yield by managing speed and power trade-offs [8, 11], and using runtime variation and process aging data to throttle chip frequency [12]. These studies have also been extended to include the sub-threshold range of supply voltages for Ultra-DVS systems [13], where the impacts of variation are shown to be even more significant [14, 15].

Self-adaptive systems use sensors to detect variations and automatically adjust frequency and voltage, though AVS systems can also be controlled by software changing the target frequency mode. Adaptive systems generally use ring oscillators [5] or critical path replicas [16, 17] to monitor the effects of voltage reduction until a target frequency is met [5, 17]; unfortunately, use of these units alone leaves a system susceptible to across-chip variation. Multiple local oscillators may be used to combat this, but with significant power and area overhead costs [18]. If the units are used to meet a frequency target and then disabled to save power, the system is also susceptible to runtime variation between mode changes. Another method for considering variation in a AVS system is the Razor technique [19–21], which detects delay errors using an extra set of clock-delayed latches, and recovers from those errors by inserting a bubble into the pipeline. These systems also incur large power and area costs. Similar to the razor technique, 'crystal ball' [22] or 'canary' [23] latches have been proposed to fail before any critical paths, so that the system parameters may be updated to avoid interrupting functionality.

Alternatively, look-up table (LUT)-based methods allow for guardbands to be set at each voltage and temperature operating point [12, 24]. Previous designs using variation tables have explored compensating for the effects of temperature variation [24], or used voltage droop detectors and temperature sensors to compensate for individual parameter fluctuations [12]. These LUT-based techniques avoid the oscillator-based design complexities of adaptive systems, and can be modified to include across-chip variations by adjusting the table values to consider worst-case process ranges; unfortunately, these worst-case considerations result in lower frequencies and higher voltages than necessary, potentially wasting a considerable amount of power and lowering yield due to power budget limits.

In this chapter, we present the individual and combined impacts of on-chip process, voltage, and temperature variations. The major design criteria for variation-tolerant, multi-voltage adaptive systems are described, and a new adaptive system is proposed combining a ring oscillator approach with a look-up table approach to improve energy efficiency. In addition, a multi-core framework is presented to illustrate how these adaptive systems may be used to control a large number of cores with a very small overhead.

4.1 Reliability Issues in Nanoscale Systems

Reliability is one of the largest challenges facing nanoscale system design. PVT variations have become increasingly problematic as technology has been scaled into the nanometer regime; at nanoscale dimensions and sub-volt supply voltages,

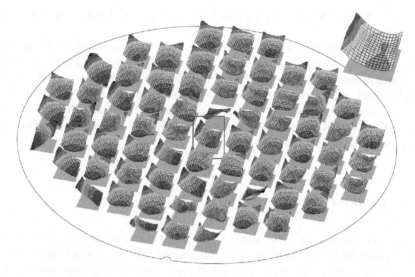

Fig. 4.1 Inter- and intra-die I_D variations in a 90 nm process [25]

even very small tolerances can result in large changes in performance and power dissipation. The increasing energy density of each new technology generation has made on-chip temperature variation critical. In addition, smaller devices have a reduced critical charge, Q_{crit}, which results in a higher probability of error resulting from particle strikes or crosstalk.

4.1.1   Process Variation

Device dimensions are now measured in the hundreds of atoms, meaning placement accuracies of even tens of atoms are not sufficient for VLSI integration. At the atomic level, quantum phenomenon (such as barrier tunneling, which has resulted in the gate leakage problems motivating the need for high-κ gate dielectrics) pose a fundamental limit to what is capable with our current fabrication methods, regardless of the accuracy of our fabrication techniques. Doping limitations are an additional quantum phenomenon which will pose a fundamental limit to scaling—if a transistor channel is only 100 atoms across, then there may only be a few dopant atoms in the entire device channel, and the placement of these dopant atoms will dramatically impact each transistor's performance. Even in the 90 nm technology node, inter- and intra-die process variations result in a large range of device performance, as shown by the variations in I_D in Fig. 4.1 [25].

Fabrication improvements such as EUV lithography [26] are currently being developed, but these techniques may not be ready for a few technology generations. In the meantime, fabrication limitations have been improved through techniques such as layout regularity [27] or double exposure [28], which are fundamental changes in design and fabrication methodologies rather than improving patterning resolutions.

Quantum limitations are more difficult to address. Some limitations, such as barrier tunneling, can be pushed back a few generations by using new materials. Other limitations have no known solution, meaning we may need to find other ways of dealing with immense on-chip process variations in future chip designs; potential solutions include local voltage control or pseudo-synchronous architectures such as globally-asynchronous locally-synchronous (GALS) or locally-asynchronous globally-synchronous (LAGS) design.

4.1.2 Runtime Variation

To minimize guardbands in the adaptive system, it is important to have a detailed model of the underlying variations being compensated. These variations are examined over the entire range of potential operating conditions to create a table of the worst-case delay impact for each condition.

Runtime variations in voltage can affect delay and power according to the well-known approximations

$$\tau = C_L V_{DD} / I_D \tag{4.1}$$

and

$$P_{dyn} + P_{sc} + P_{leak} = \alpha_{sw} C_L V_{DD} V_{swing} f + V_{DD} t_{sc} I_{peak} f + V_{DD} I_{leak} \tag{4.2}$$

where τ_d is delay, C_L is load capacitance, V_{DD} is supply voltage, I_D is drain current, P_{dyn} is dynamic power dissipation, P_{sc} is short-circuit power dissipation, P_{leak} is leakage power dissipation, α_{sw} is switching activity factor, V_{swing} is voltage swing, f is clock frequency, t_{sc} is the time short circuit power is dissipated, I_{peak} is the peak current drawn during switching, and I_{leak} is leakage current.

Temperature variation also affects these metrics through device mobility and threshold voltage, as mentioned in Chap. 2.

To examine the combined effects of runtime variation, drain currents (I_D) of equally-sized, diode-connected, bulk NMOS and PMOS devices from the 90 nm BSIM4 predictive technology model [29] are examined (the plot of I_D vs. voltage vs. temperature is shown in Fig. 2.6). The operating conditions chosen include temperatures between $-55°C$ and $125°C$ and voltages between 200 mV and 1.2 V. The changing I_D trends in the $+I$-T and $-I$-T slope regions cause the impact of each type of variation to change depending upon the operating voltage. For example, a $\pm 12°C$ temperature variation at 0.4 V causes almost no change in current, while the same variation at 1.2 V can cause a large change in current. The combined delay impacts of voltage and temperature variation at each operating point are used to create frequency guardbands. The individual effects of each variation are also

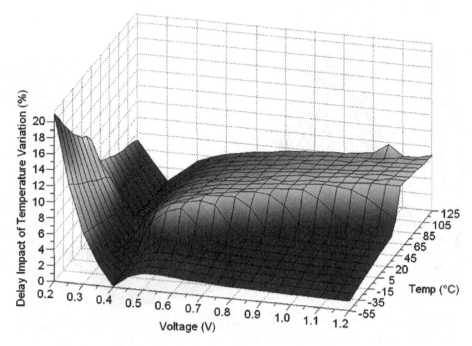

Fig. 4.2 Inverter delay sensitivity to ±12°C temperature variation

useful for comparing different guardband techniques and are examined in the following subsections.

4.1.3 Temperature Variation Impact

The reversal of the temperature dependence at low voltages is caused by the temperature, voltage, and technology dependencies of threshold voltage and mobility [30–32]. To determine the effect of temperature on delay performance, an on-chip temperature variation model is taken from a chip and package thermal model of an IC [33]. In this model, temperature varies by 24°C across the chip under normal conditions. Figure 4.2 shows the worst-case delay impact due to a ±12°C variation on a symmetrically-sized inverter at each operating point (for example, the 1.2 V, 80°C operating point shows the change in delay caused by temperature between 68°C and 92°C).

Two areas of note in Fig. 4.2 are the temperature insensitive valley between 350 and 400 mV (recall that V_{INS} in the 90 nm technology occurs in this region), and the two distinct regions at higher voltages caused by the effect of temperature on symmetric delay sizing. These non-linear behaviors show the importance of measuring variation impacts with respect to specific operating conditions rather than designing an adaptive system around fixed temperature thresholds.

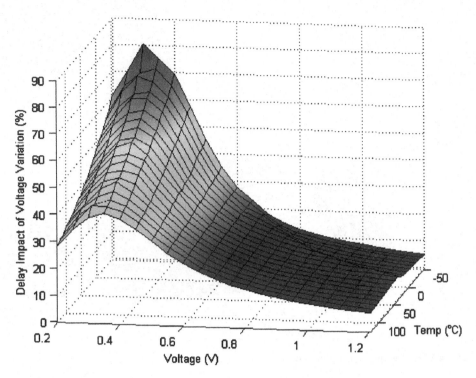

Fig. 4.3 Inverter delay sensitivity to ±10% supply voltage variation

4.1.4 Voltage Variation Impact

Two major contributors to supply voltage variation are fluctuation in the supply load, which can change by up to 700% depending upon input patterns [2], and IR drop, caused by non-zero interconnect resistance. Variations in supply voltage can also arise from substrate noise or interconnect coupling. Reduction of voltage variations to 2% V_{DD} have been reported using decoupling capacitors [34]; however, supply variations up to 10% V_{DD} are common, even with well-designed distribution networks and decoupling capacitors [7].

Using a ±10% fluctuation in supply voltage [7], Fig. 4.3 shows the impact of voltage variation on delay across the same range of voltages and temperatures from Fig. 4.2. At extremely low voltages, the ±10% voltage range becomes small enough that the associated change in current begins to taper off slightly, though the delay impact is still approximately three times that at nominal voltage. Comparing Figs. 4.2 and 4.3, voltage variation clearly has a much larger delay influence than temperature variation at low voltages, while delay impacts are comparable in the nominal voltage region.

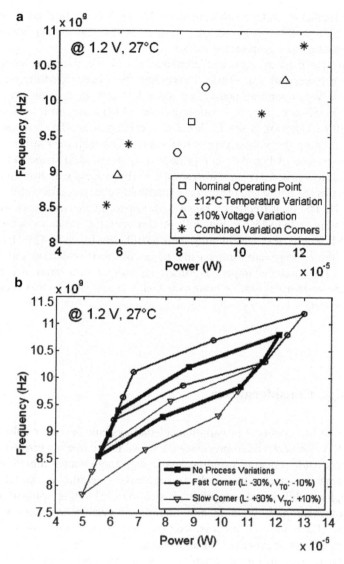

Fig. 4.4 Runtime variation impacts (**a**) with no process variation and (**b**) with fast and slow process corners

4.1.5   Combined Effects of Variation

The combined effects of runtime variation are shown in Fig. 4.4a for an 11-stage ring oscillator at 1.2 V and 27°C. The square marker shows the performance of the system prior to the introduction of variation; the circular and triangular markers show the performance impacts of the temperature and voltage variations, respectively. As shown, the impact of temperature variation on frequency is nearly as large as the impact of voltage variation at the 1.2 V/27°C nominal point (expected from Figs. 4.2 and 4.3), while voltage variation causes a far greater fluctuation in power dissipation than temperature variation.

The four star markers represent combinations of the variation sources. For example, the lower left star marker represents the system performance at the +12°C/−10% V_{DD} operating point. The lower left and upper right star markers indicate the maximum range of frequencies over which a system must function to tolerate runtime variation at the 1.2 V/27°C operating point. To show the importance of accounting for process variation in conjunction with the runtime variation analysis, the outline of Fig. 4.4a is represented by the boldest shape in Fig. 4.4b, with fast and slow process corners also shown. In the process variation model, gate length, L, varies by ±30% and the primary threshold voltage coefficient, V_{TO}, varies by ±10% [7]. The total frequency range with respect to the worst case increases from 26% in Fig. 4.4a to 45% in Fig. 4.4b. At this operating point, existing look-up table methods would be overestimating their guardbands by up to 45%. Even more concerning, these ranges are further widened as voltage is scaled, as shown in the next section. The combined impact of voltage and temperature variation is recorded at each of the system's operating points for both fast and slow process corners for later comparison.

4.2   Design Considerations

The metrics to be optimized in variation-tolerant systems depend on the application. In high performance microprocessors, the goal is to surpass aggressive performance targets while reducing variation just enough to guarantee functionality [8]. Low performance applications such as pacemakers require a careful balance between high reliability and low power dissipation [35]. Other applications are a mix of these examples, yielding any number of performance and variability requirements. Each form of process, voltage, and temperature variation has its own challenges. While process [36] and voltage [37] variations are both on-chip forms of variation, temperature variability comes from both on-chip and environmental sources. In order to better control the effects of temperature variability, it should be included in the design process along with speed and power performance targets.

Supply voltage is among the most important design decisions because of its control of a large trade-off between speed and power [38, 39]. In nanoscale technologies, choice of supply voltage also gives a degree of control over temperature variation depending on the gate overdrive, V_{GS}-V_T. Combining this control with the voltage dependence of speed ($T \propto CV/I$) and power ($P \propto CV^2f$) performance, requirements for each application space can be mapped to a range of supply voltages.

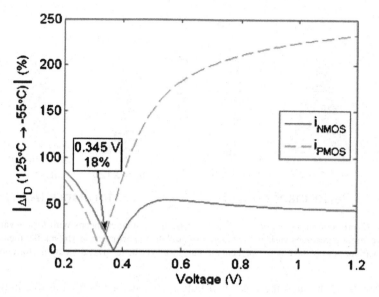

Fig. 4.5 Percentage-based variation design method

4.2.1 Percentage-Based Design for Variation

To improve robustness, designs can be selected based on the percentage change a variation causes at a specific operating point [40]. Figure 4.5 shows the percent variation in I_D caused by temperature fluctuations across a range of supply voltages (the y-axis is an absolute value, so the inflection points occur at the transition between the normal and reverse temperature dependence regions). As shown, the combined optimal point, 0.345 V, can reduce temperature variation by a factor of 12 vs. nominal voltage.

While this operating point may be useful for sensors with severe temperature robustness requirements, in most applications a purely percentage-based design decision is a gross overestimate of actual requirements. This is because a percentage-based technique completely ignores the context which is often most important—the effect on performance. Operating at 0.345 V vs. 1.20 V may reduce temperature variation by a factor of 10; however, this ultra-low voltage operation also reduces the maximum operating speed by a factor of almost 20, which limits the usefulness of the optimization with respect to large-scale system design.

Figure 4.6 shows a separate but equally important concern with percentage-based design—the additional weight given to designs with larger mean performance values. In Fig. 4.6a, delay distributions for two arbitrary designs are provided, where the distribution on the left has a smaller mean and twice the variance of the distribution on the right. Using a percentage-based design guideline, the right distribution would be chosen due to its smaller standard deviation from the mean; unlike Fig. 4.5 however, this figure shows the underlying context of performance.

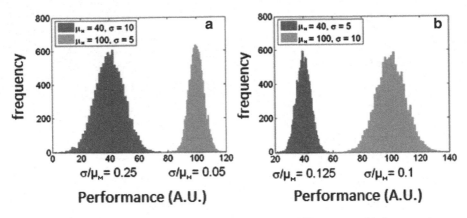

Fig. 4.6 Overly conservative percentage-based design (a) smaller means with *larger* variances may have larger percentage variation, yet better overall delay performance, (b) smaller means with *smaller* variances may have larger percentage variation, yet better overall delay performance

In this context, although the right distribution has smaller σ, the left distribution is shown to offer better performance than the right distribution—even its rightmost tail has a smaller delay than the right distribution, making it more appealing for most systems (assuming the wider distribution does not cause race conditions).

While a case can be made for both distributions in Fig. 4.6a, b examines an even more concerning situation for percentage-based design. Here, the left distribution has a smaller mean and a smaller variation, yet a design based on percent variation will indicate that the right distribution is more desirable because the σ/μ_m percentage of the right distribution is smaller than that of the left distribution. This illustrates the previously mentioned skew towards designs with larger μ_m, an important drawback of design comparisons using constant ratios.

4.2.2 Yield-Based Design for Variation

Another commonly used design technique is to optimize for some pre-selected performance limit [8]. In this case, variations negatively impacting performance and causing a design to miss the yield cutoff are considered losses (but may be binned as a slower corner). Whereas percentage-based design for variation largely ignores the context of performance, yield-based design for variation ignores the differences between design distributions. Figure 4.7 shows two closely related distributions for an arbitrary performance metric, with the star distribution having a slightly larger mean and slightly smaller variance than the circle distribution. Yield-based design cutoffs are shown by the vertical lines in Fig. 4.7, with each line indicating a performance-based yield cut-off (or different binning levels of the same design).

In this example, the desired yield cut-off determines the appropriate design choice. From Table 4.1, the solid line at the performance value of 60 includes

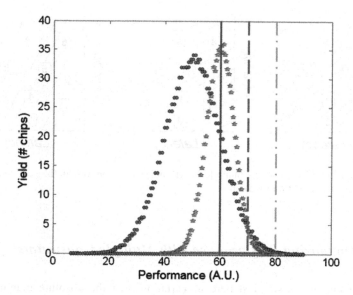

Fig. 4.7 Yield-based variation design method

Table 4.1 Yield percentages associated with Fig. 4.7

Yield	60 ▬▬	70 ▬ ▬	80 ▬ ı ▬
●	84.205%	97.717%	99.869%
★	50.025%	97.718%	99.998%

84% of the circle distribution and only 50% of the star distribution, making the circle distribution the obvious choice; however, the dashed line at 70 is much less conclusive, with 97.7% of each distribution included within the limit. The dash-dotted line at 80 illustrates the third case, with the smaller tail of the star distribution making it the proper choice.

Yield is a function of a large number of design variables, with some options having better average performance with large potential variations and other options having less variation at the cost of lower performance. Without taking into account the distribution of each design decision, yield-based design can far from optimal—simply choosing the best performance in each decision may not be the best use of resources. Instead, decisions with large performance differences (e.g. a 1% delay improvement resulting in a 5% power increase) should be saved until all other decisions have been made. By doing this, the remaining decisions can be viewed in a larger context, allowing small concessions in one decision to yield large benefits in another, and resulting in a global solution with larger safety margins.

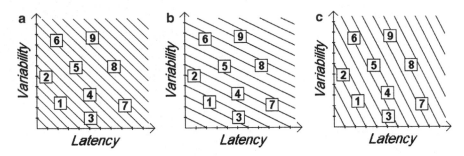

Fig. 4.8 Variability vs. latency for multiple design choices (**a**) even weight, (**b**) weighted towards variability, (**c**) weighted towards latency

4.2.3   Optimizing Performance with Multiple Constraints

Instead of taking the smallest percent variation over the absolute magnitude or setting a performance limit for each component, a combination of the two schemes can optimize robustness together with the other types of performance. This combination provides a more efficient alternative to the iterative approach of tweaking performance and robustness requirements one at a time, and also allows application-specific weighting of the necessary metrics. Fig. 4.8a shows an application with equal weighting between latency and variability, with the sloped lines marking unit 1 as the optimal choice. Figure 4.8b mimics a sensor network application, where variability is weighted more heavily and unit 3 could be used in place of unit 1. Figure 4.8c shows a high-performance application, weighted towards latency, where unit 2 would be optimal.

By assigning weight coefficients to the performance and variability needs of the system, each design choice can be mapped in terms of both variation and performance. A simple minimization function then provides the optimal decisions through each step in system design, allowing for trade-offs to be made to meet hard latency or variability limits.

4.2.4   Temperature-Robust Performance Yield through Supply Voltage Selection

To combine delay, power, and temperature variation, supply voltages are mapped to a set of target frequencies by including the effect of temperature variation in the delay performance at each supply voltage. Using a datapath of six FO4 inverters in a 90 nm technology, a Monte Carlo analysis of temperature variation is performed by modeling temperature as a normal Gaussian distribution [41]. On-chip temperature variation is taken from a full-chip and package thermal model of an IC [33]

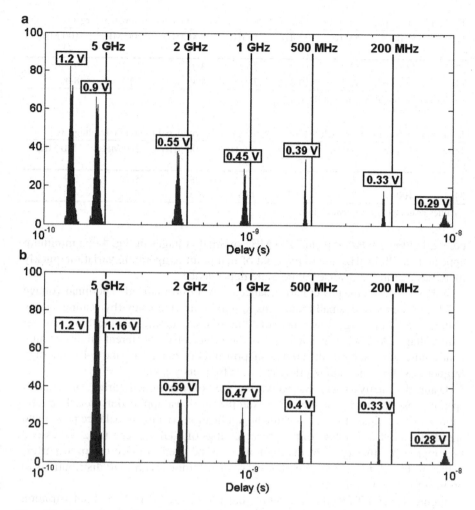

Fig. 4.9 Delay distributions of optimal supply voltages including temperature variation for different performance target (**a**) 27°C mean, (**b**) 74°C mean

whose temperature varies by 24°C under optimal conditions. This 24°C variation is used to set a 3σ of ± 12°C. Each delay distribution is generated by translating a 5,000-point temperature distribution about a mean, with σ/μ_m values generated from each data set. Two environmental operating points are modeled via distributions about 27°C and 74°C means to extend this analysis to a range of applications.

   Figure 4.9 shows a number of performance goals and the associated optimal voltage selections (with 10 mV precision) at each mean temperature. Note that in Fig. 4.9b, the nominal and optimal voltage distributions overlap. Including variability in the supply voltage selection allows for supply voltage to be scaled to the appropriate performance goal while also considering the impact on variation, numerically shown in Table 4.2. Trends can also be graphically observed by

Table 4.2 σ/μ_m percentage for the optimal supply voltage at each performance target

Temp (°C)	5 GHz* (%)	5 GHz (%)	2 GHz (%)	1 GHz (%)	500 MHz (%)	200 MHz (%)	100 MHz (%)
27	13.8	13.7	10.8	7.7	3.4	3.4	9.2
74	13.7	13.6	10.7	7.7	4.1	2.1	8.0

*Values collected at nominal voltage

Table 4.3 Power used for each optimal supply voltage normalized to nominal voltage at 74°C

Temp (°C)	5 GHz* (%)	5 GHz (%)	2 GHz (%)	1 GHz (%)	500 MHz (%)	200 MHz (%)	100 MHz (%)
27	98.1	50.6	17.4	10.7	7.8	5.4	3.9
74	100	92.3	19.9	11.9	8.3	5.4	4.0

*Values collected at nominal voltage

comparing the 5 GHz, 200, and 100 MHz optimal voltages in Fig. 4.9; a minimum appears near 200 MHz, and to the right of that point temperature variation quickly increases due to the reverse temperature dependence.

Table 4.3 shows the power performance analysis, normalized to nominal voltage at 74°C. As expected, small reductions in maximum frequency (by scaling supply voltage) can yield large power benefits due to the quadratic relationship of power with voltage. Also seen in Table 4.3 is the effect of the different environmental conditions, with a large difference in power between 27°C and 74°C at high frequencies, but a negligible effect at lower frequency points.

Using this analysis with the weighting method mentioned earlier, application specific designs can be targeted. For example, a robust application would target a supply voltage closer to 0.33 V, while an application wishing to balance power and speed might take advantage of the power savings of 2 GHz over 5 GHz. To extend this approach, other types of distributions may be used in place of speed performance in Fig. 4.9 (e.g. replacing the delay distributions with PDP distributions to optimize low-power designs).

Figure 4.9 and Table 4.2 also show that use of the 0.345 V optimal variation voltage requires an unnecessary performance hit with respect to nominal voltage. In that range, a small increase in supply voltage can provide large speed gains while only giving up a small amount of robustness and power performance. Looking to the right of that optimal variation voltage, a large reduction in frequency is required to achieve modest gains in power or variability.

4.3   Adaptive Delay Correction in Adaptive Voltage Scaling Systems

This section examines the impact of runtime variations over the multiple operating voltages of an AVS system. Two popular methods of variability compensation adaptively vary supply voltage and/or body bias [8, 9, 12, 16, 17]. Prior scaling

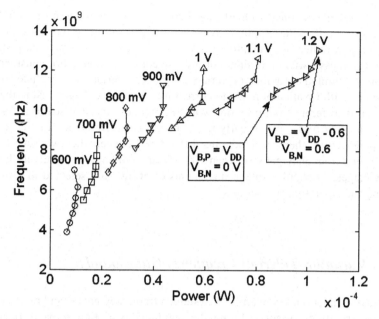

Fig. 4.10 Impact of adaptive supply voltage and adaptive body bias on an 11-stage ring oscillator

methods are briefly examined to determine appropriate operating voltages for the AVS analysis. This work shows how a change in the variation sensor can improve an adaptive system's power efficiency while ensuring robustness. In the proposed system, voltages are adjusted to improve power consumption, not to counter the effects of variation.

4.3.1 Defining the Operating Voltages

Previous work comparing variable supply and body bias schemes has determined that either is sufficient to compensate for variation [8]—process variation was the specific type chosen in that work—with the combination of the two schemes yielding marginal improvement. To examine the effects of the adaptive schemes on runtime variation, an 11-stage ring oscillator was simulated using the 90 nm BSIM4 predictive technology model [29] at supply voltages between 0.6 and 1.2 V and body biases between standard body bias (SBB) and 0.6 V of forward body bias (FBB). Beyond 0.6 V FBB, device functionality is limited by gate leakage. The maximum frequency and average power of each operating point are plotted in Fig. 4.10. As shown, adjusting supply voltage has a much greater impact on power and energy dissipation than body bias—reducing V_{DD} from 1.2 to 1 V at 0.6 V FBB provides the same frequency as lowering the bias voltage to 0.5 V FBB, while offering over $10\times$ the power savings.

Supply voltage scaling is clearly the better option for reducing power in an adaptive system, thus the proposed dynamic voltage scaling system uses adaptive supply voltage rather than adaptive body bias. Adaptive body bias may still be useful for process variation tuning, but process compensation is not considered in this work to limit the number of variables (process variation is included in the guardbands, with no attempt made to correct it using bias voltages). Note that this conclusion is different from previous work [8] which found that scaling of supply voltages or body biases have nearly the same effect on frequency tuning; from a power perspective, there is a definite difference between the two schemes. In Fig. 4.10, the frequency gaps between 100 mV are shown to be relatively large at lower voltages, so supply voltage increments of 50 mV are selected to provide a finer range of operating frequencies.

4.3.2 Variation-Tolerant Frequency Guardbands

To guarantee functionality in an AVS system, a frequency guardband must be used to tolerate runtime variation. Using the combined variation impacts presented earlier in this chapter, the system is limited by the worst-case temperature variation at each voltage; however, by separating the impact into voltage and temperature components, each component can be individually adjusted depending on the maximum expected variation over a period of time, resulting in an overall reduced guardband. Because temperature varies relatively slowly, on the order of 10 μs/°K [42], the temperature variation component of the guardband can be nearly eliminated if the system can adapt quickly enough to take into account changes in the temperature before system functionality is affected. Unfortunately, supply voltage can spike and droop on a sub-nanosecond timescale [43], meaning that little can be done to adapt to the fastest harmonics aside from reducing the magnitude of these variations. The separation of voltage and temperature variation can be implemented using voltage [44] and temperature sensor readings.

The total guardband requirement, $f_{guardband}$, is defined by the maximum frequency changes due to voltage and temperature variation before the adaptive system is able to respond.

$$f_{guardband} \geqslant \left(\frac{\partial f}{\partial V} \cdot \frac{dV}{dt} + \frac{\partial f}{\partial T} \cdot \frac{dT}{dt} \right) \cdot t_{response} \qquad (4.3)$$

$$t_{response} = t_{sensor} + t_{correction} \qquad (4.4)$$

$\partial f / \partial V$ and $\partial f / \partial T$ are the frequency impact of voltage and temperature variation, respectively; dV/dt and dT/dt are the voltage droop and temperature change rates, respectively. The adaptive system response time, $t_{response}$, is the sum of the polling time of the temperature and voltage sensors, t_{sensor}, and the time it takes to

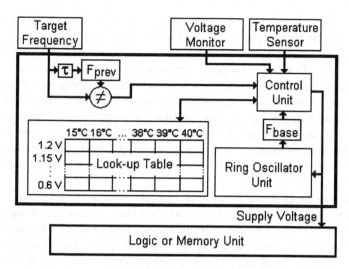

Fig. 4.11 Proposed variation-tolerant system

adaptively correct any detected changes, $t_{correction}$. Parameters t_{sensor}, $t_{correction}$, dV/dt, and dT/dt, are all implementation-specific. For example, voltage droop rates depend on the nature of power distribution networks [2], and the frequency correction latency may be reduced by using multiple Phase-Locked Loops (PLLs) [12]. For reference, reported $t_{response}$ latencies are 0.3 ns [12], and t_{sensor} delays are 1.4 ns [44] for voltage sensors and 50 ns for temperature sensors, depending upon the desired sensor resolution.

4.3.3   Proposed Adaptive System Design

The proposed system design is shown in Fig. 4.11. On start-up, the control unit takes a frequency value from the ring oscillator unit, which is stored as a baseline frequency, f_{base}, for that voltage. f_{base} is updated with each change in voltage to include the voltage dependency of process effects. Once f_{base} is determined, the control unit polls the voltage and temperature sensors to determine the starting operating point. This data is then passed as an address to the LUT, which returns the percent overhead (the guardband) that is required to ensure voltage and temperature robustness. This percent overhead is combined with f_{base} to calculate $f_{guardband}$.

This result, $f_{guardband}$, is the minimum operating frequency required to ensure correct functionality at that voltage. If the target frequency input is faster than $f_{guardband}$, the control unit increases the voltage so that the controlled circuit will operate at a faster frequency. This new voltage affects the ring oscillator frequency, which is passed back into the control unit to generate a new, higher $f_{guardband}$. The control unit will keep increasing the voltage until $f_{guardband}$ is faster than the external target frequency, to ensure robustness. Note that for any increase in target frequency, the adaptive system will need time to guarantee an appropriate guardband before that frequency can be applied to the system.

The system will also decrease the voltage to save power when lower frequencies are acceptable. If the external target frequency is much slower than $f_{guardband}$, then the control unit will decrease the voltage until the two frequencies fall within a pre-determined threshold of the target operating frequency. This threshold is important to ensure that $f_{guardband}$ never drops below the target frequency, which could cause functional failure. Because the guardband is stored as a percentage of f_{base}, and not fixed to a slow process variation corner, any frequency gain resulting from a faster process corner can be used to further reduce the voltage and improve power savings.

Three inputs are used in the control unit to determine if a new voltage is required to maintain robustness. A new voltage can be selected because of a change in target frequency, or because of changes in the voltage and temperature sensors. While temperature generally varies on a very slow timeframe, low-power idle modes may also run at very slow frequencies, making temperature variation guardbands more important. To prevent errors due to voltage fluctuations, voltage minima over a given time window should be recorded using a real-time sampling method [44].

A flowchart illustrating the method just described is provided in Fig. 4.12. After the control unit calculates the starting operating point using the ring oscillator and LUT, the system begins its polling loop of the three inputs. Accessing the appropriate LUT entries, the robustness loop guarantees that the guardbands are followed to preserve functionality.

After robustness is ensured, changes in voltage and temperature operating points may enable lower supply voltages to be used if these voltages meet the guardband requirement at the target frequency. This is handled in the low-power loop. The separation of the two loops is important, as the speed of the robustness loop is vital to both system functionality and the amount of power savings which can be achieved (how much the guardband can be reduced while avoiding the potential for timing failure).

An example of the look-up table design and operation is provided in Fig. 4.13, showing the organization of the voltage and temperature operating points along with an example scenario highlighting the system functionality. For the purposes of explanation, frequency numbers are used instead of percentages; the actual data in the lookup table are based on a percent overhead, not a frequency magnitude, to make use of the ring oscillator baseline.

4.3.4 Results and Comparison with Prior Work

In Table 4.4, the temperature and voltage variation models are used to create a complete list of runtime variation effects at each voltage under fast, nominal, and slow process corners. The four datasets shown include the following four parameters: the operating frequency of the ring oscillator without variation,

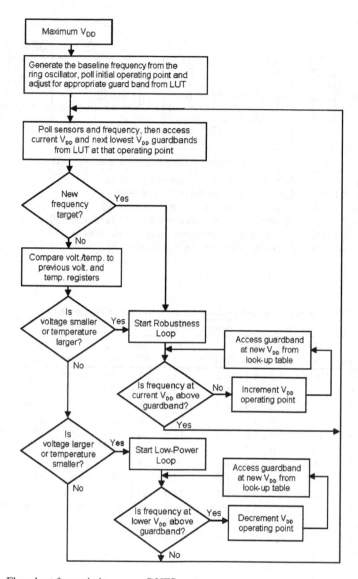

Fig. 4.12 Flowchart for variation-aware DVFS system

f (GHz); worst case combined delay changes from the previously described runtime variation models, W.C.; and the individual contributions of the voltage variation component, ΔV, and temperature variation component, ΔT.

The temperature contribution is shown to be relatively constant across all process corners, while the impact of voltage variation has a more noticeable change with process variation, also increasing dramatically as voltage decreases. These results will be further analyzed to compare existing look-up table methods with the proposed method.

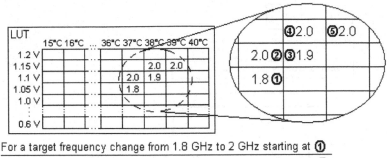

For a target frequency change from 1.8 GHz to 2 GHz starting at ①

Step 1) Location ① does not meet requirement, move to location ②
Step 2) Location ② meets requirement, but temperature has
increased due to the higher voltage, move to location ③
Step 3) Location ③ does not meet requirement, move to location ④
Step 4) Location ④ meets the requirement, and the increase in
temperature has reached a steady state at location ⑤

Fig. 4.13 Example of look-up table operation

Table 4.4 Delay impact of runtime variation in a ring oscillator for multiple process corners

	No process variation				Slow corner				Fast corner			
		W.C.	ΔV	ΔT		W.C.	ΔV	ΔT		W.C.	ΔV	ΔT
VDD	f(GHz)	(%)	(%)	(%)	f(GHz)	(%)	(%)	(%)	f(GHz)	(%)	(%)	(%)
1.20	9.71	10.7	5.9	4.7	9.10	12.3	7.1	5.1	10.30	9.2	4.7	4.4
1.15	9.40	11.4	6.5	4.8	8.76	13.2	7.8	5.2	10.01	9.8	5.3	4.4
1.10	9.06	12.2	7.2	4.9	8.39	14.2	8.7	5.3	9.70	10.5	5.9	4.5
1.05	8.68	13.2	8.0	4.9	7.99	15.2	9.6	5.4	9.36	11.3	6.6	4.6
1.00	8.28	14.2	8.9	5.0	7.56	16.5	10.7	5.5	8.98	12.2	7.4	4.6
0.95	7.83	15.4	10.0	5.2	7.09	17.9	11.9	5.7	8.57	13.2	8.2	4.7
0.90	7.35	16.8	11.2	5.3	6.59	19.6	13.4	5.8	8.11	14.4	9.3	4.8
0.85	6.83	18.4	12.6	5.4	6.05	21.5	15.1	5.9	7.62	15.7	10.5	4.9
0.80	6.26	20.3	14.3	5.5	5.46	23.9	17.1	6.1	7.07	17.3	11.8	5.0
0.75	5.65	22.7	16.4	5.6	4.84	26.9	19.8	6.2	6.48	19.1	13.5	5.2
0.70	4.98	25.6	18.9	5.8	4.17	30.8	23.2	6.3	5.83	21.4	15.5	5.2
0.65	4.28	29.3	22.3	5.8	3.46	35.8	27.8	6.3	5.13	24.2	18.1	5.3
0.60	3.52	34.3	27.0	5.8	2.73	42.7	34.3	6.1	4.38	27.9	21.4	5.3

As mentioned, voltage change can be quite fast compared to changes in temperature. Because of this, the ΔV column percentages indicated in Table 4.4 include the entire range of voltage variations from the variation model presented earlier in this chapter. The temperature variation is slower, and the system can be designed to adjust to the temperature variation as it changes, and take advantage of the increased speed at lower temperatures (or higher temperatures in the reverse dependence region) to further reduce voltage and power; however, to take into account the potential for on-chip temperature variation, the ΔT column is also taken over the entire ±12°C temperature range. If the regions with widely varying temperature (for example the core and cache) were on different voltage islands, each could be individually adjusted to reduce the guardbands and save power.

To compare the proposed system with other approaches, an accurate way of determining the minimum voltage requirements based on fluctuations in the voltage/temperature operating point is needed. Thus, an adaptive simulator was created to take into account the effect of power dissipation on temperature. As seen in Fig. 4.4, large changes in temperature cause only small changes in power, so the system converges to a steady-state temperature and power dissipation at each target frequency (the potential for thermal runaway is not included in this work). The simulator provides a temperature, voltage, and frequency-dependent power analysis of the adaptive system, using extracted simulation results from the unit under test to ensure accuracy. The modeled system is capable of dynamically scaling V_{DD} between 1.2 and 0.6 V in steps of 50 mV. Multiple guardband methods are compared using a sample program which runs through a set of operating frequencies evenly distributed between 300 and 800 MHz, mimicking a DVFS system. The power calculation is iterative; that is, the first frequency is input into the simulator at a nominal condition, with the temperature at each subsequent point calculated using the following power-dependent temperature model.

Runtime temperature changes are modeled iteratively using the common junction temperature formula [45]

$$T_j(t) = T_a + P(t) \cdot R_{th} \qquad (4.5)$$

where T_j and T_a are the junction and ambient temperatures, respectively, P is the power dissipated at that iteration's operating point, and R_{th} is the thermal resistance (°C/W) [46]

$$R_{th} = \frac{1}{2\pi k_{th}} \left[\frac{1}{L_x} \ln\left(\frac{L_x + \sqrt{W_x^2 + L_x^2}}{-L_x + \sqrt{W_x^2 + L_x^2}} \right) + \frac{1}{W_x} \ln\left(\frac{W_x + \sqrt{W_x^2 + L_x^2}}{-W_x + \sqrt{W_x^2 + L_x^2}} \right) \right] \qquad (4.6)$$

In (4.6), k_{th} is the thermal conductivity of silicon (0.15 mW/μm°C) and L_x and W_x are treated as the dimensions of the unit under test, allowing the entire unit to be treated as a variable heat source. The unit under test was fit to the temperature model by calculating R_{th} and then scaling P by a constant such that the maximum power dissipation resulted in a 24°C rise above ambient temperature. To avoid inaccuracies caused by rapid thermal changes, it is assumed that time spent at each frequency target is large with respect to the thermal time constant. This simplifies the results by allowing the steady-state temperature to be used to calculate the power dissipation at each frequency target. Otherwise, the system power would be dependent on the previous frequency target as well as the current frequency target (the transient temperature changes depend on the difference in power dissipation between the frequency targets). A more detailed thermal behavior model could be used for more accuracy [46], but for the desired first-order analysis that level of detail is unnecessary.

Table 4.5 Look-up table comparison for a typical corner chip

VDD	LUT-only method W.C. (GHz)	Proposed method W.C. (GHz)	% increase in frequency
1.20	7.98	8.51	6.74
1.10	7.20	7.77	7.94
1.00	6.31	6.91	9.49
0.90	5.30	5.91	11.57
0.80	4.16	4.76	14.58
0.70	2.89	3.45	19.59
0.60	1.56	2.02	29.12

Table 4.6 Dynamic guardband power savings assuming a worst-case process corner

		Worst case process		Proposed method	
Method:		W.C.	W.C./V	W.C.	W.C./V
Temp. (°C):		±12	±12	±12	±12
Volt. (%):		±10	±10	±10	±10
Freq. (MHz)	800	1.00	0.66	1.00	0.52
	700	1.00	0.57	0.81	0.50
	600	1.00	0.62	0.80	0.54
	500	1.00	0.68	0.78	0.59
	400	1.00	0.75	0.75	0.64
	300	1.00	0.85	0.72	0.72
Avg.		1.00	0.69	0.81	0.58

Table 4.5 shows the increase in frequency achieved by the proposed system with the baseline frequency and percentage-based look-up table compared to a worst-case PVT look-up table approach on a typical corner chip. In the LUT-only column, $f_{guardband}$ is calculated by multiplying the slow corner percentages from Table 4.4 by the slow corner operating frequencies. For the proposed approach, $f_{guardband}$ is calculated by multiplying the same percentages by f_{base}, which corresponds to the typical corner operating frequencies. As shown, the proposed method offers significant frequency advantages at each voltage, which will translate into increased power savings in the adaptive system. Note that the percent frequency increase grows significantly as supply voltage decreases, indicating that dynamic voltage and frequency scaling (DVFS) systems operating at low and ultra-low voltages would benefit even more from this approach.

The look-up table can be populated in a number of ways, depending on the size of the table and the specific variation models used. Table 4.6 shows the power results of the worst case method versus the proposed method for two different guardband implementations. Power savings in the controlled unit are reported across a 300–800 MHz frequency range. A four-stage ripple carry adder was used to compare the effects of the systems; four stages were the maximum number achieving the 800 MHz frequency with a worst-case guardband in a 90 nm technology. The results are normalized to the overall worst case (W.C.) system in the worst case process implementation.

In Table 4.6, the W.C. system represents a DVFS system with a constant guard band across the entire range of operating points (meaning no adaptive control is required). It is the smallest possible look-up table implementation, only requiring one entry, and uses the largest guardband from Table 4.4, which corresponds to the 0.6 V supply voltage. The W.C./V system includes the combined variation results presented earlier to dynamically adjust the guardband at each supply voltage, which requires a larger table and matches the approach described in Fig. 4.13.

As shown, the proposed method reduces the total power consumption by 20% in the W.C. system and by 15% in the W.C./V system, resulting in a 42% power reduction in the proposed W.C./V system over a design without any adaptive control. These results show that there is a clear benefit to using the ring oscillator to generate a frequency baseline with a percentage-based look-up table.

4.4 Multi-Core System Design with Variation Tolerance and a Low Power Safety Mode

All of the aforementioned techniques guarantee reliability at the expense of large power and complexity overheads. In this section, we focus on a much simpler safety mode module with a small power and area footprint, allowing it to be duplicated as necessary to ensure reliability in a multi-core system. The safety mode module determines when a core is near delay failure and reduces that core's frequency by half, ensuring correct functionality even under extreme worst-case operating conditions.

4.4.1 Multi-Core Framework

An important component of any adaptive system that compensates for variation is a sensor to track the runtime operating conditions. Depending on the application environment, this sensor (or multiple sensors, if necessary) may be required to track a wide variety of effects, such as PVT variation, power dissipation, or substrate noise. The sensors must have some way of quantifying their results to determine when and how to adjust the adaptive system to maintain functionality. One such framework is shown in Fig. 4.14.

The safety mode module presented in this section is capable of halving the frequency in each core individually whenever runtime variations approach levels which could cause system failure, in this case due to delay error. By limiting the per-core overhead to a simple frequency divider and one column of a LUT (the failure-threshold vector of the sensor output), this framework is extendable to a large number of cores which would be unmanageable by systems with large per-core overhead requirements such as PLLs or voltage regulators. The purpose of the

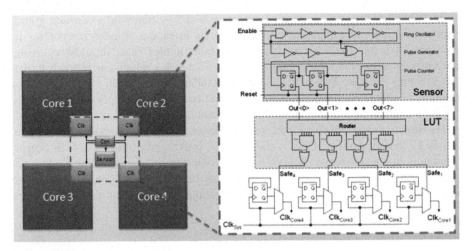

Fig. 4.14 Multi-core framework for variation-tolerant design

safety mode module is variation tolerance, not power optimization, and the large frequency margin ensures correct functionality over a wide range of conditions. Future work will examine the potential for these systems to be used in automatic load leveling (using the arbiter to give priority to the fastest operating cores), as well as power and area trade-offs associated with more than two choices of operating frequency.

The example multi-core framework shown in Fig. 4.14 includes a sensor, look-up table (LUT), four computation cores, an arbiter for handling the I/O of each core, and system memory. While the arbiter controls the flow of information, the sensor and LUT control the arbiter and core frequencies. Figure 4.14 also provides a closer look at the safety mode module. As shown, the sensor feeds its quantized information to the LUT, which is created at design time (the LUT function can be implemented using simple gates, as shown, or an addressable array of memory elements with a comparator). The LUT controls whether or not to implement the safety mode in each core using the *Safe* signals, which choose from a system clock and the output of a frequency divider. The LUT also provides this information to the arbiter, which controls the data rate to each core.

The sensor used for the test case, shown in Fig. 4.14, is an all-digital delay sensor [47], modified by including a pulse generator (the XOR gate fed by a two inverter buffer) to double the sensor resolution and make it more sensitive to runtime changes in temperature. The sensor can also be used to measure die-to-die process variation, or voltage losses due to IR drop, by recording the sensor output at a specific temperature and comparing that result to other known values (similar to how performance-sensitive ring oscillators are used [48]). The accuracy of the all-digital sensor is not limited by process mismatch and DC offset as in analog sensors [49], though it is susceptible to voltage variations due to changing loads. These voltage variations must be built into the frequency guardband of each unit.

Voltage spikes are also problematic due to their short duration, which could result in oscillation between the normal and safety modes. This oscillation could be reduced by adding hysteresis to the system, for example using an incrementer to artificially increase the digital output when the safety mode is triggered, which would avoid this oscillation pattern until the temperature was reduced enough to reset the hysteresis signal.

The sensor is composed of three parts, shown in Fig. 4.14. The delay sensing mechanism is an enable-controlled ring oscillator, which will continue to run as long as the enable signal is asserted. By adjusting the pulse width of the enable signal, very small variations in delay can be accumulated across multiple clock cycles using the pulse counter. The oscillator is tapped and fed into a pulse generator, which converts every transition in the ring oscillator into a full pulse, effectively doubling the sensor accuracy. The pulse generator feeds a pulse counter, which tallies up the number of pulses and provides a digital read-out, with each temperature corresponding to an n-bit state. n is determined by the maximum number of pulses the sensor will count, a function of the enable pulse width and the range of operating conditions. If necessary, the capacity of the pulse counter can be doubled by simply adding a flip-flop to the end of the chain. For the example application, n is set to 8 bits, and the sensor can achieve a resolution of 4°C when enabled for 30 ns. Reduced sensor accuracies require the safety mode to be asserted at lower temperatures (meaning larger temperature safety margins are needed), at the expense of lost system throughput.

4.4.2   Multi-Core Simulation Results

Schematic level simulations were performed in Cadence Spectre using a 90 nm technology model [29]. Each core in this example system is a 64-bit adder composed of cascaded 4-bit carry lookahead blocks. To simulate a more general system where each core could have different frequency requirements and variation tolerances, each core is given a different body bias. Core 1 is given a normal bias ($V_{B,P} = 1$ V, $V_{B,N} = 0$ V), Core 2 is given 0.2 V of forward bias ($V_{B,P} = 0.8$ V, $V_{B,N} = 0.2$ V), Core 3 is given 0.4 V of forward bias ($V_{B,P} = 0.6$ V, $V_{B,N} = 0.4$ V), and Core 4 is given 0.6 V of forward bias ($V_{B,P} = 0.4$ V, $V_{B,N} = 0.6$ V).

To illustrate the impact of variation on the system, the temperature response of the four cores is shown with and without process and voltage variation in Fig. 4.15. As shown, the combination of process and voltage variation can result in nearly 3× delay degradation at 125°C, and increases the temperature dependence of the delay in each core. Below 0°C, the delay in Core 4 begins to taper off. This may be caused by a change in the temperature dependence of the core (the threshold voltage is affected by body bias, and threshold voltage is one of the major temperature dependence controllers). The impact of body bias on the temperature dependence will be examined in more detail in future work.

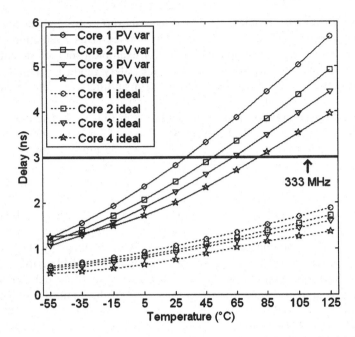

Fig. 4.15 Temperature response of the four cores with and without process and voltage variation

Figure 4.15 is also useful as a visualization tool for safety mode operation, as illustrated by a simple example. Drawing a horizontal line at 333 MHz as shown, it is possible to find the temperatures at which each core can no longer operate at that frequency. When process and voltage variation are considered, we can see that Core 1 will fail beyond ~30°C, Core 2 will fail at ~47°C, Core 3 will fail at ~64°C, and Core 4 will fail near 80°C. Applying the safety mode module to these cores reduces the operating frequency of each core by 2× (changing the clock period from 3 to 6 ns) when a core approaches its failure point. From Fig. 4.15, we see that all cores have a delay of <6 ns up to 125°C; thus, using the safety mode system, all four cores are variation-tolerant across the entire range of military specified temperature conditions, albeit with reduced throughput beyond those temperature limits. Without this adaptive capability, system frequency would be limited to the worst case frequency of 167 MHz.

The safety mode system operation is shown in Fig. 4.16 at 27°C with 10% voltage variation and a worst-case process corner. As described earlier, the *Enable* signal in the temperature sensor is asserted over a number of clock cycles to accumulate the delay error in the ring oscillator. After that pulse has completed, the *Readout* signal passes the sensor output to the LUT. Beyond 27°C, Core 1 is incapable of operating correctly at the 333 MHz system clock, so the Safe 1 signal is activated, indicating Core 1 needs to be put into the low-frequency safety mode.

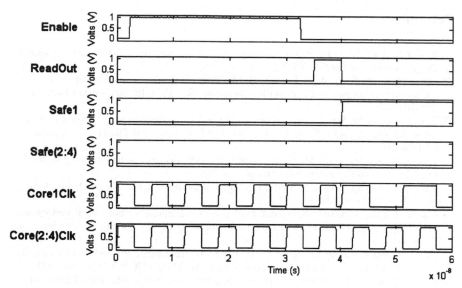

Fig. 4.16 Safety mode system operation at 27°C

Core 2, Core 3, and Core 4 are all capable of operating correctly at 333 MHz, so the other Safe signals are kept at logic low. The safety mode is shown to operate correctly on the Core1Clk signal, with the operating frequency dropping from 333 to 167 MHz after the 40 ns system latency. The clock signals to the other three cores remain unchanged. In future designs, the first cycle of the safety mode clock will be skipped to avoid the partial clock pulse during transition between modes, shown in Fig. 4.16 at 40 ns on signal Core1Clk.

For our particular system and technology, the power overhead of the safety mode module, including the sensor, LUT, and clock dividers, is 105.1 μW per adaptation at 27°C; however, because temperature changes very slowly (on the order of microseconds) the sensor is left inactive for the majority of runtime, making its power overhead very small compared to the 638.1 μW average power of the combined cores (also measured at 27°C). The total latency of adaptation is 40 ns, including the sensor latency, LUT and clock dividers. Assuming one sample is taken every 10 μs [49], the average power of the safety mode module is just 0.56 μW. Idle leakage power in the module is 0.14 μW; thus, this average could be further reduced using a power gating scheme. The area overhead of the safety mode module (including the sensor and clock adjustment circuitry in each core) is only 3% of the overall multi-core system, making our approach both power- and area-efficient for variation-tolerant multi-core design. This small power and area overhead would also allow multiple sensors to be implemented as necessary to take into account intra-die variation.

References

1. Borkar S et al (2003) Parameter variations and impact on circuits and microarchitecture. Design Automation Conf 338–342
2. James N, Restle P, Friedrich J, Huott B, McCredie B (2007) Comparison of split- versus connected-core supplies in the POWER6 microprocessor. IEEE Int Solid-State Circuits Conf 298–604
3. Wu Q, Pedram M, Wu X (2000) Clock-gating and its application to low power design of sequential circuits. IEEE Trans Circuits and Syst I: Fundamental Theory and Applications 47:415–420
4. Jiang H, Marek-Sadowska M, Nassif S (2005) Benefits and costs of power-gating technique. IEEE Int Conf on Computer-Aided Design 559–566
5. Rabaey J, Chandrakasan A, Nikolic B (2003) Digital Integrated Circuits: A Design Perspective, 2nd ed. Prentice Hall, NJ
6. Cao Y, Clark L (2005) Mapping statistical process variations toward circuit performance variability: an analytical modeling approach. ACM IEEE 42nd Design Automation Conf 658–663
7. Cao Y, Gupta P, Kahng AB, Sylvester D, Yang J (2002) Design sensitivities to variability: extrapolations and assessments in nanometer VLSI. IEEE Int ASIC/SOC Conf 411–415
8. Chen T, Naffziger S (2003) Comparison of adaptive body bias (ABB) and adaptive supply voltage (ASV) for improving delay and leakage under the presence of process variation. IEEE Trans Very Large Scale Integr (VLSI) Syst 5:888–899
9. Martin S, Flautner K, Mudge T, Blaauw D (2002) Combined dynamic voltage scaling and adaptive body biasing for lower power microprocessors under dynamic workloads. IEEE/ACM Int Conf on Computer-Aided Design 721–725
10. Shakeri K, Meindl J (2002) Temperature variable supply voltage for power reduction. IEEE Comp Soc Ann Symp on VLSI 64–67
11. Tschanz J, Narendra S, Nair R, De V (2003) Effectiveness of adaptive supply voltage and body bias for reducing impact of parameter variations in low power and high performance microprocessors. IEEE J Solid-State Circuits 38:826–829
12. Tschanz J, et al (2007) Adaptive frequency and biasing techniques for tolerance to dynamic temperature-voltage variations and aging. IEEE Int Solid-State Circuits Conf 292–604
13. Calhoun B, Chandrakasan A (2006) Ultra-dynamic voltage scaling (UDVS) using sub-threshold operation and local voltage dithering. IEEE J Solid-State Circuits 41:792–804
14. Hanson S et al (2006) Ultralow-voltage, minimum-energy CMOS. IBM J Res and Dev 50:469–490
15. Zhai B, Hanson S, Blaauw D, Sylvester D (2006) Analysis and mitigation of variability in subthreshold design. Int Symp Low Power Electronics and Design 20–25
16. Elgebaly M, Sachdev M (2007) Variation-aware adaptive voltage scaling system. IEEE Trans Very Large Scale Integr (VLSI) Syst 15:560–571
17. Kao J, Miyazaki M, Chandrakasan A (2002) A 175-mV multiple-accumulate unit using an adaptive supply voltage and body bias architecture. IEEE J Solid-State Circuits 37:1545–1554
18. Zhang K, Nguwen D, Lenehan D (2005) Method and apparatus for providing supply voltages for a processor. US Patent 6948079
19. Das S et al (2006) A self-tuning DVS processor using delay-error detection and correction. IEEE J Solid-State Circuits 41:792–804
20. Lee S et al (2004) Reducing pipeline energy demands with local DVS and dynamic retiming. Int Symp Low Power Electronics and Design 319–324
21. Ernst D et al (2004) Razor: circuit-level correction of timing errors for low-power operation. IEEE Micro 24:10–20
22. Eireiner M, Henzler S, Georgakos G, Berthold J, Schmitt-Landsiedel D (2007) In-situ delay characterization and local supply voltage adjustment for compensation of local parametric variations. IEEE J Solid-State Circuits 42:1583–1592

23. Calhoun B, Chandrakasan A (2004) Standby power reduction using dynamic voltage scaling and canary flip-flop structures. IEEE J Solid-State Circuits 39:1504–1511
24. Gyohten T et al (2006) An on-chip supply-voltage control system considering PVT variations for worst-caseless lower voltage SoC design. IEICE Trans on Electronics E89-C:1519–1525
25. Masuda H, Ohkawa SI, Kurokawa A, Aoki M (2005) Challenge: variability characterization and modeling for 65- to 90-nm processes. IEEE Custom Integrated Circuits Conf 593–599
26. Silverman PJ (2002) The Intel lithography roadmap. Intel Tech J 6:55–61
27. Jhaveri T et al (2007) Maximization of layout printability/manufacturability by extreme layout regularity. J Micro/Nanolith MEMS MOEMS 6:1–15
28. Fritze M et al (2003) Enhanced resolution for future fabrication. IEEE Circuits and Devices Mag 19:43–47
29. Zhao W, Cao Y (2006) New generation of predictive technology model for sub-45 nm early design exploration. IEEE Trans Electron Devices 53:2816–2823
30. Filanovsky IM, Allam A (2001) Mutual compensation of mobility and threshold voltage temperature effects with applications in CMOS circuits. IEEE Trans Circuits and Syst I: Fundamental Theory and Applications 48:876–884
31. Park C et al (1995) Reversal of temperature dependence of integrated circuits operating at very low voltages. Int Electron Devices Mtg 71–74
32. Bellaouar A, Fridi A, Elmasry MI, Itoh K (1998) Supply voltage scaling for temperature-insensitive CMOS circuit operation. IEEE Trans Circuits and Syst II: Analog and Digital Signal Processing 45:415–417
33. Huang W, Humenay E, Skadron K, Stan M (2005) The need for a full-chip and package thermal model for thermally optimized IC designs. Int Symp Low Power Electronics and Design 245–250
34. Ji G, Arabi T, Taylor G (2005) Design and validation of a power supply noise reduction technique. IEEE Trans Adv Packaging 28:445–448
35. Maheshwari A, Burleson W, Tessier R (2002) Trading off reliability and power-consumption in ultra-low power systems. Int Symp Quality Electronic Design 361–366
36. Multu A, Gunther N, Rahman M (2003) Concurrent optimization of process dependent variations in different circuit performance measures. IEEE Int Symp Circuits and Syst 692–695
37. Bobba S, Thorp T, Aingaran K, Liu D (2001) IC power distribution challenges. IEEE/ACM Int Conf Computer-Aided Design 643–650
38. Soeleman H, Roy K, Paul B (2001) Robust subthreshold logic for ultra-low power operation. IEEE Trans Very Large Scale Integr (VLSI) Syst 9:90–99.
39. Wang A, Chandrakasan A, Kosonocky S (2002) Optimal supply and threshold scaling for subthreshold CMOS circuits. IEEE Comp Soc Ann Symp VLSI 5–9
40. Kumar R, Kursun V (2006) Impact of temperature fluctuations on circuit characteristics in 180 nm and 65 nm CMOS technologies. IEEE Int Symp Circuits and Syst 3858–3861
41. Ajami A, Banerjee K, Pedram M (2005) Modeling and analysis of nonuniform substrate temperature effects on global ULSI interconnects. IEEE Trans Computer-Aided Design of Integrated Circuits 24:849–861
42. Semenov O, Vassighi A, Sachdev M (2006) Impact of self-heating effect on long-term reliability and performance degradation in CMOS circuits. IEEE Trans Device and Materials Reliability 6:17–27
43. Muhtaroglu A, Taylor G, Rahal-Arabi T (2004) On-die droop detector for analog sensing of power supply noise. IEEE J Solid-State Circuits 39:651–660
44. Sato T et al (2007) On-die supply-voltage noise sensor with real-time sampling mode for low-power processor applications. IEEE Int Solid-State Circuits Conf 290–603
45. Tadyon P (2000) Thermal challenges during microprocessor testing. Intel Tech J 1–8
46. Rinaldi N (2001) On the modeling of the transient thermal behavior of semiconductor devices. IEEE Trans Electron Devices 48:2796–2802

47. Griffith S (1989) Method and apparatus for measuring the speed of an integrated circuit device. US Patent 4890270
48. Xiong J, Zolotov V, Visweswariah C, Habitz PA (2008) Optimal margin computation for at-speed test. Design, Automation and Test in Europe 622–627
49. Luh L Jr, Choma J, Draper J, Chiueh H (1999) A high-speed CMOS on-chip temperature sensor. IEEE European Solid-State Circuits Conf 290–293

Chapter 5
Controlling the Temperature Dependence

Temperature variations affect system speed, power, and reliability by altering the threshold voltage (V_T), mobility (μ), and saturation velocity (v_{sat}) in each device [1, 2]. The resulting changes in device current can cause a number of reliability issues such as delay uncertainty. The extent of the temperature variations depends on the application; for example, on-chip temperature gradients of up to 50°C have been reported [3], while the US military requires ICs to function over a wide range of environmental conditions, from −55°C to 125°C [4].

The impact of temperature variation on delay and power varies with supply voltage, shown in Fig. 5.1 for an inverter in a commercial 65 nm technology. In Fig. 5.1a, the delay uncertainty over the −55°C to 125°C temperature range is shown by the error bars at each supply voltage. An inverter at 1 V has a delay uncertainty of 14.5%; however, the uncertainty is heavily dependent upon the application—as shown in Table 5.1, a 3 mm interconnect link may have delay uncertainty in excess of 200% at 0.6 V [5].

Another important observation from Table 5.1 is that the delay change from −55°C to 125°C is negative at 0.6 V and positive at 1.0 V. The voltage region where delay increases with temperature is referred to as the normal temperature dependence region, while the region where delay decreases with temperature is referred to as the reverse (or inverted) temperature dependence region [6]. Between the two regions, there is a supply voltage where the impact of temperature dependence on delay is minimized [7–9]. This is referred to as the temperature-insensitive voltage V_{INS}, and as technology scales this voltage approaches nominal voltages [2, 9].

V_{INS} may be exploited to improve temperature resiliency and performance stability, reduce frequency guardband requirements and clock skew, or enable more precise measurements of other types of variation. Unfortunately, use of V_{INS} restricts design to a very specific delay and power operating point, limiting the versatility of the V_{INS} approach. In addition, NMOS and PMOS devices have different V_{INS} values [9], limiting the achievable temperature resilience.

Note that these temperature insensitive approaches and those proposed in this chapter are not intended to improve circuit delay; indeed, $V_{INS} = 710$ mV for an svt

D. Wolpert and P. Ampadu, *Managing Temperature Effects in Nanoscale Adaptive Systems*, DOI 10.1007/978-1-4614-0748-5_5, © Springer Science+Business Media, LLC 2012

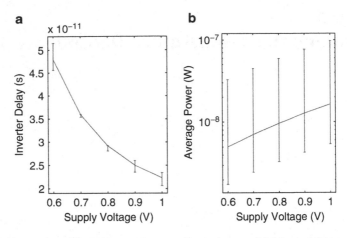

Fig. 5.1 Error bars showing the impact of temperature on inverter (**a**) delay and (**b**) average power in a 65 nm technology over a range of supply voltages

Table 5.1 Temperature-induced delay uncertainty in a 65 nm technology

	Delay uncertainty over military temperature range				
Test unit	0.6 V	0.7 V	0.8 V	0.9 V	1.0 V
FO4 inverter	12.9%[a]	1.4%[a]	5.7%	10.6%	14.5%
3 mm link [5]	210.0%[a]	59.3%[a]	8.2%[a]	11.4%	19.3%

[a] Delay at −55°C larger than delay at 125°C

device in a 65 nm technology, resulting in ~50% larger delays than operation at nominal voltage. Instead, these methods are exploited to improve circuit reliability for application-critical systems, or ensure proper functionality of supporting circuits such as clock trees and on-chip sensors.

In this chapter, we present a method that reduces delay uncertainty more effectively than merely exploiting V_{INS}, enabling temperature resilience over a wide voltage range. The remainder of this chapter is organized as follows: Sect. 5.2 investigates existing V_{INS}-based temperature-insensitive methods, Sect. 5.3 presents the proposed temperature-insensitive methodology, Sect. 5.4 provides a set of example applications for the proposed methodology, and Sect. 5.5 discusses limitations of the proposed approach.

5.1 Existing Methods for Reducing Temperature Sensitivity

V_{INS} has been used to improve the temperature resilience of a variety of logic circuits [10] and to improve clock skew [11]. Multi-V_T design [12] and adaptive body biasing [13] have been proposed to adjust V_{INS}, enabling a range of design targets to achieve low temperature sensitivity; however, the small degree of V_{INS}

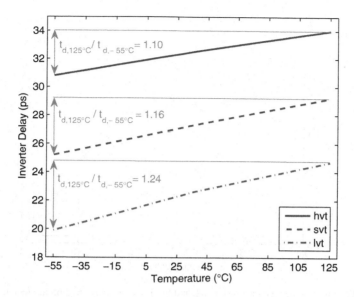

Fig. 5.2 Impact of temperature on lvt, svt, and hvt inverter delays in a 65 nm technology for $V_{DD} = 1$ V, with temperature sensitivities $t_{d,125°C}/t_{d,-55°C}$ labeled

tuning they achieve (~100 mV) still limits operation to points near the technology-specific values.

These V_{INS} techniques operate at a fixed bias regardless of changes in temperature. To improve temperature resiliency at other voltages, temperature-adaptive alternatives such as thermal throttles [14] or safety modes [15] may be used. These systems use temperature sensors to detect operating conditions and dynamically adjust system parameters at runtime to avoid temperature-induced timing failures and other reliability issues. These adaptive methods are more complex than V_{INS} designs, and require multiple temperature sensors, adaptive circuits, and integrated control systems.

5.1.1 Temperature Insensitivity with Multi-Threshold Design

The impact of temperature on inverter delay is a function of V_T; thus, low-threshold voltage (lvt), standard-threshold voltage (svt), and high-threshold voltage (hvt) devices each have different temperature dependencies, as shown in Fig. 5.2. Delay changes monotonically with temperature (aside from a slight parabolic behavior when the dependence is reduced below ~1%); thus, we will refer to the temperature dependence of delay as the ratio of the delays at 125°C and −55°C, $t_{d,125°C}/t_{d,-55°C}$.

V_{INS} occurs at a single supply voltage for a given threshold voltage. Thus, the use of multiple threshold voltages enables designers to take advantage of multiple

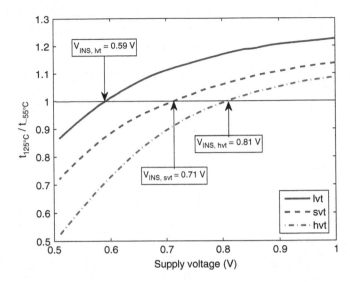

Fig. 5.3 Temperature dependence of lvt, svt, and hvt inverter delays in a 65 nm technology with V_{INS} labeled

values of V_{INS} [12]. The values of V_{INS} $(t_{d,125°C}/t_{d,-55°C} = 1)$ for lvt, svt, and hvt devices in a 65 nm technology are shown in Fig. 5.3; y-axis values >1 represent the normal temperature dependence region, while values <1 indicate the reverse temperature dependence region. Lower threshold devices are shown to have a V_{INS} at lower supply voltages, as expected from the analysis in Chap. 2. Note that the temperature dependence decreases sharply at low voltages; thus, using lvt instead of svt at low voltages will have a much larger impact on temperature resilience than using hvt instead of svt at high voltages.

Multi-threshold designs have also been used effectively when lvt devices are in the reverse temperature dependence region and hvt devices are in the normal dependence region; in these conditions, proper interleaving of these devices in a datapath can reduce temperature sensitivity and reduce power [12, 16].

5.1.2 Temperature Insensitivity with Adaptive Body Biasing (ABB)

Whereas single-V_T design offers one value of V_{INS} and multi-V_T design offers three discrete V_{INS} to choose from, adaptive body biasing (ABB)—modifying the source-body voltage V_{SB}—offers a continuous range of V_T values to choose from, adjusting V_T as shown in (5.1) [17].

$$V_T(T) = V_{T,V_{B0}} + \gamma\left(\sqrt{|-2\phi_F(T) + V_{SB}|} - \sqrt{|-2\phi_F(T)|}\right) \qquad (5.1)$$

where

$$V_{T,V_{B0}} = \Phi_{GC}(T) - 2\phi_F(T) - \frac{Q_{B0}(T)}{C_{ox}} - \frac{Q_{ox}}{C_{ox}} \tag{5.2}$$

$$\gamma = \frac{\sqrt{2q \cdot N_A \cdot \varepsilon}}{C_{ox}} \tag{5.3}$$

$$\phi_F(T) = \frac{kT}{q} \ln \frac{N_A}{n_i(T)} \tag{5.4}$$

$V_{T,VB0}$ is the unbiased threshold voltage, γ is the body-effect coefficient, and ϕ_F is the Fermi potential. $V_{T,VB0}$ and ϕ_F each depend on temperature (γ has a slight temperature dependence because of the permittivity ε [18], but it is small and generally ignored), as shown in (5.2)–(5.4). In (5.2), Φ_{GC} is the gate-channel work function difference, Q_{B0} is the depletion region charge density at surface inversion, and Q_{ox} is the oxide-substrate interface charge density. Φ_{GC} and Q_{B0} are temperature dependent because each have ϕ_F terms [17]. In (5.3) and (5.4), q is a unit charge, N_A is the dopant concentration, and n_i is the intrinsic carrier concentration (also a function of T [17]). Simply put, V_{SB} adjusts V_T, altering the overall MOSFET temperature sensitivity as described in Chap. 2.

The impact of adaptive body biasing on an svt inverter's temperature dependence is shown in Fig. 5.4. The x-axis values in Fig. 5.4 are the NMOS bias, V_{BN}—the PMOS bias V_{BP} is equal to $V_{DD} - V_{BN}$. Each curve in Fig. 5.4 represents a different supply voltage, with $V_{DD} = 0.7$ shown to intersect the $t_{d,125°C}/t_{d,-55°C} = 1$ line in Fig. 5.4a, indicating the delay is nearly insensitive to temperature for $V_{BN} = -0.1$ V. As shown in Fig. 5.4b, increasing V_{BN} beyond ~0.6 V causes a dramatic increase in power consumption, making large-scale integration at these biases infeasible. Using $V_{BN} < 0.6$ V, body biasing can achieve $t_{d,125°C}/t_{d,-55°C} = 1$ in svt inverters for 0.625 V $< V_{DD} < 0.75$ V.

ABB can also be combined with multi-V_T designs; however, each device type (lvt, svt, hvt) has different temperature dependences, requiring individual body biases per device type to achieve insensitivity, resulting in very large overheads.

5.2 Proposed Approach for Temperature Insensitivity

While prior approaches manipulate V_{INS} by controlling V_T, those approaches are limited by the range of V_T tuning we can achieve. The use of adaptive body biasing and/or multi-V_T design methodologies can provide temperature insensitivity for supply voltages between ~0.6 and ~0.8 V; however, nominal voltage in the 65 nm technology used in this chapter is 1.0 V.

We propose a new circuit technique capable of reducing temperature insensitivity to within ~1% over a wide range of supply voltages (0.6–1.0 V) with a single

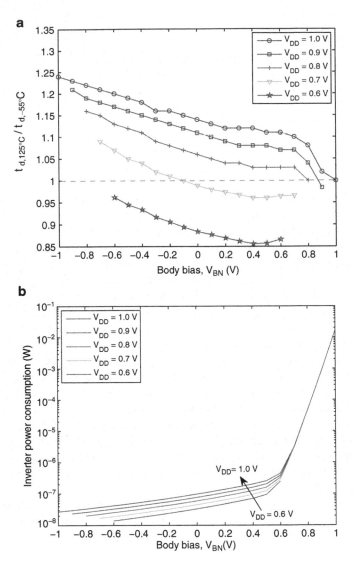

Fig. 5.4 Adaptive body biasing an svt inverter. (**a**) Impact on temperature sensitivity, (**b**) impact on power consumption

device type (svt); thus, the proposed approach is a more versatile alternative to the V_{INS}, multi-V_T, and ABB approaches.

Temperature insensitivity is achieved by adjusting the gate overdrive voltage ($[V_{GS} - V_T]$, from the device current model in Chap. 2). In prior approaches, this is achieved by adjusting V_T or V_{DD} (V_{GS} is normally equal to V_{DD}). In this work, instead of controlling $[V_{GS} - V_T]$ through V_{DD} or V_T, we propose to control $[V_{GS} - V_T]$ by controlling V_{GS} a different way. The insertion of an additional device with a programmable gate voltage allows us to tune V_{GS} without having to change V_{DD},

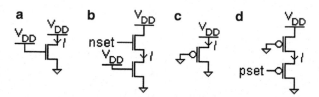

Fig. 5.5 Schematics used to generate Fig. 5.6. (**a**) Changing V_{DD} for NMOS devices, (**b**) changing V_{nset} for NMOS devices, (**c**) changing V_{DD} for PMOS devices, (**d**) changing V_{pset} for PMOS devices

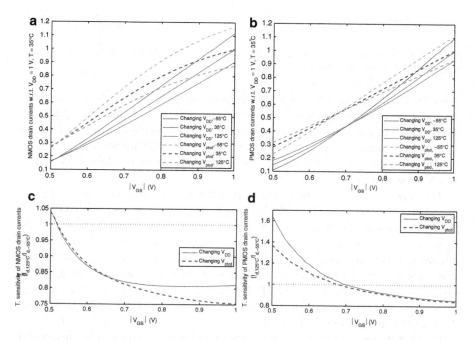

Fig. 5.6 Comparison of changing V_{DD} to adjust temperature sensitivity and adjusting V_{PTCD} to adjust temperature sensitivity. (**a**) Normalized NMOS drain currents, (**b**) normalized PMOS drain currents, (**c**) ratio of current at 125°C and −55°C in NMOS device, (**d**) ratio of current at 125°C and −55°C in PMOS device

using the structures shown in Fig. 5.5b, d. We name these additional devices programmable temperature compensation devices (PTCDs), and they are the basis of our new temperature insensitive circuit technique.

Controlling V_{GS} through V_{pset} and V_{nset} allows us to achieve a similar impact on the temperature dependence of I_D as controlling V_{GS} by scaling V_{DD}, as shown in Fig. 5.6. Figure 5.6a compares the schematics shown in Fig. 5.5a, b. The solid lines in Fig. 5.6a show the impact of changing V_{DD} in Fig. 5.5a (where $|V_{GS}| = V_{DD}$), while the dashed lines in Fig. 5.6a show the impact of changing the gate voltage *nset* in Fig. 5.5b while keeping $V_{DD} = 1$ V (in this case, $|V_{GS}| = V_{nset}$). As shown, the two behaviors both achieve temperature insensitivity around $|V_{GS}| \approx 0.51$ V; however,

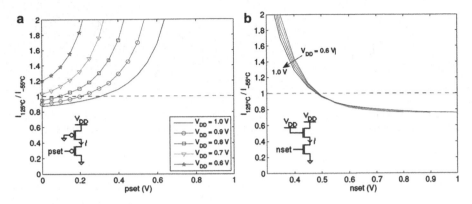

Fig. 5.7 Temperature insensitivity achieved in (**a**) PUN and (**b**) PDN using programmable temperature compensation structure

using the PTCD we have achieved this temperature insensitivity while maintaining V_{DD} at 1 V (nominal voltage)! Thus, with the PTCDs we are able to achieve temperature insensitivity at any voltage between nominal voltage and the technology's 'nominal V_{INS}.' In this technology, the PMOS device V_{INS} ($V_{INS,P}$) is 0.71 V, so we are able to achieve temperature insensitive operation for 0.71 V < V_{DD} < 1 V, as shown in Fig. 5.7.

The y-axes in Fig. 5.7 indicate the ratio of the network currents (labeled I in the insets) at 125°C ($I_{125°C}$) and −55°C ($I_{−55°C}$). Figure 5.7a was generated by computing the current through two series-connected PMOS devices (shown in the figure)—one with source connected to V_{DD} and gate connected to ground, and the other with drain connected to ground and gate biased to a voltage *pset* (the x-axis). The setup for Fig. 5.7b is similar to Fig. 5.7a using NMOS devices and a voltage *nset*, also shown in the figure. The PMOS curves in Fig. 5.7a are shown to cross $I_{125°C}/I_{−55°C}$ = 1 (indicated temperature insensitivity) down to $V_{DD} \approx 0.7$ V, while the NMOS curves in Fig. 5.7b cross $I_{125°C}/I_{−55°C}$ = 1 down to V_{DD} = 0.6 V.

In Fig. 5.8, we examine how to integrate PTCDs into a temperature insensitive system design, summarizing the major trade-offs of the proposed technique. On the left we show the conventional design using a supply voltage of V_{INS}. The insensitivity achieved by this approach is limited by the difference in $V_{INS,P}$ and $V_{INS,N}$. To achieve temperature insensitivity, a designer can use the average of the PMOS and NMOS V_{INS} values or set the supply voltage to $V_{INS,P}$ ($V_{INS,N}$) and size the pull-up network PUN (or pull-down network PDN) to dominate the worst-case delay over the entire temperature range, achieving 'worst-case temperature insensitivity'.

Including a single PTCD, shown in the center of Fig. 5.8, allows for the circuit to operate at a supply voltage of $V_{INS,P}$ while adjusting the *nset* device gate voltage to make the PDN temperature-insensitive at that supply. This achieves better temperature-insensitivity at the cost of a power-delay product ($P*\tau$) penalty from adding the PTCD (the improvements are quantified in Table 5.2 for a single inverter in an 11-stage ring oscillator using 2:1 P/N ratios for the conventional $V_{INS,P}$ design and 1:1 P/N ratios for the *PTCD* designs, which minimized $P*\tau$) Note that in this design we are still restricted to operation at a single supply voltage, $V_{INS,P}$.

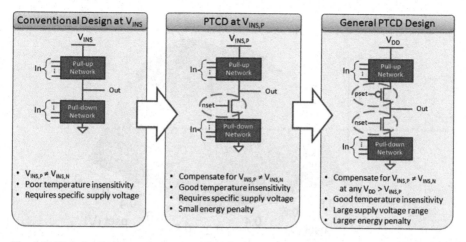

Fig. 5.8 Techniques for employing programmable temperature compensation devices in a conventional static CMOS logic gate

Table 5.2 Delay and power comparison of temperature insensitivity methods for 65 nm inverter

Design	Supply voltage	Worst case delay (ps)	Avg. power @ 27°C	P-τ product (aJ)	Delay uncert. (%)
$V_{INS,P}$	710 mV	14.5	606 nW	8.8	12.9
$PTCD_{nset}$	710 mV	28.7	715 nW	20.5	0.9
$PTCD_{nset+pset}$	1 V	43.2	1.17 μW	50.5	0.3

The most versatile approach involves adding two PTCDs—one below the PUN and one above the PDN. In this approach, we can adjust *pset* and *nset* to achieve temperature insensitivity at any supply voltage larger than $V_{INS,P}$, although there is a larger energy penalty associated with the addition of two extra devices into the gate. Note that for the PTCD designs, it is critical to place the PTCD adjacent to the output; if the PTCD is placed adjacent to the source, it will create a virtual supply node that will not achieve the desired temperature sensitivity benefits.

For either PTCD approach, the P/N ratio does not affect the temperature sensitivity like in the conventional V_{INS} design; instead, *pset* and *nset* control the temperature sensitivity while the P/N ratio determines the relationship between the rising and falling edge delays.

5.2.1   PTCD Inverter

The impacts of the *nset* and *pset* gate biases on the temperature sensitivity of a PTCD inverter at 1 V are shown in Fig. 5.9. The x-axis in Fig. 5.9a indicates the *nset* voltage, the y-axis indicates the *pset* voltage, and the z-axis indicates the impact of

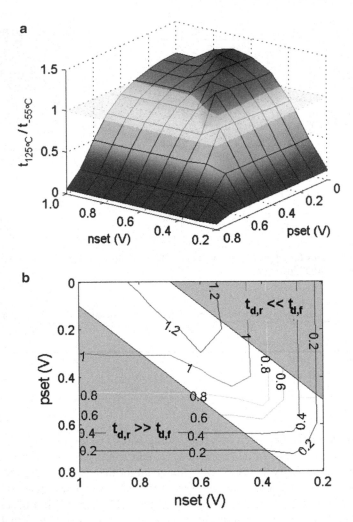

Fig. 5.9 (**a**) Surface plot of PTCD inverter temperature dependence at nominal voltage with z = 1 plane showing temperature-insensitive biases. (**b**) Contour plot of temperature dependence with worst-case edge delays labeled

temperature on the worst-case delay over the −55°C to 125°C range ($t_{125°C}/t_{−55°C}$). The intersecting plane at z = 1 indicates temperature-insensitive bias points (where $t_{d,125°C} = t_{d,−55°C}$). When z > 1, the inverter delay has a normal temperature dependence; when z < 1, the delay has a reverse temperature dependence.

The surface data in Fig. 5.9a is flattened into the contour plot in Fig. 5.9b to show the relationship between the rising and falling edge delays on the temperature sensitivity of the worst-case delay. When the rising edge delay $t_{d,r}$ is the worst-case delay, changes in *nset* do not affect the worst-case temperature sensitivity, resulting in the horizontal lines in the lower left portion of Fig. 5.9b; in this region,

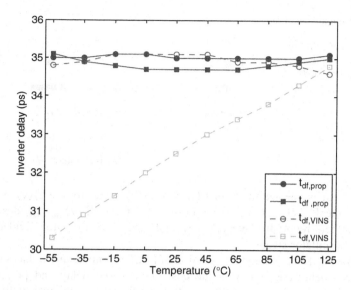

Fig. 5.10 Programmable threshold voltage structure achieves low sensitivity for both rising *and* falling edge delays. V_{DD} = 710 mV

a *pset* value of ~0.3 V makes the rising edge delay insensitive to temperature, closely matching the $I_{125°C}/I_{-55°C}$ = 1 point from Fig. 5.7a. Similarly, when the falling edge delay $t_{d,f}$ is the worst-case delay, changes in *pset* do not affect the worst-case temperature sensitivity, resulting in the vertical lines in the upper right portion of Fig. 5.9b; in this region, *nset* ≈ 0.45 V achieves temperature insensitivity.

The region where the rising and falling edges intersect is irregular because one edge is the worst case delay at low temperatures while the other edge is the worst case delay at high temperatures. When *pset* = 315 mV and *nset* = 440 mV, the delay uncertainty of both edge delays is reduced to ~1% over the temperature range, shown by the solid lines in Fig. 5.10.

The independent tuning of the PUN and PDN provide large improvements in the overall delay uncertainty over prior approaches. As shown in Fig. 5.10, the prior V_{INS} approach has a rising edge delay uncertainty $t_{dr,VINS}$ of just 0.9% over the temperature range, yet the falling edge delay uncertainty $t_{df,VINS}$ is 12.9% (this is because the NMOS and PMOS devices have different values of V_{INS} [9]). In contrast, the proposed methodology's rising edge delay uncertainty $t_{dr,prop}$ is reduced by 66% compared to $t_{dr,VINS}$, while the falling edge delay uncertainty $t_{df,prop}$ is reduced by 93% compared to $t_{df,VINS}$.

The three approaches described in Fig. 5.8 are compared qualitatively in Table 5.2. We report the worst case delay over the entire temperature range, the average power consumption at room temperature for two performance targets, and the delay uncertainty over the entire temperature range. To ensure a realistic slew rate (which can also affect temperature sensitivity as will be shown momentarily), this data was collected using ring oscillators of each gate type. To ensure a fair power comparison, we adjusted the number of stages in each oscillator to achieve

Fig. 5.11 Comparison of the voltage ranges achieving temperature insensitivity for the single-V_T, multi-V_T, ABB, and proposed methods

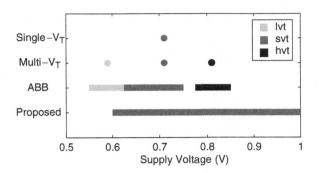

oscillator frequencies within 4% of 1 GHz, using 37 stages for the $V_{INS,P}$ design, 17 stages for the $PTCD_{nset}$ design, and 11 stages for the $PTCD_{pset+nset}$ design. The power numbers report the average power of the entire oscillator divided by the number of oscillator stages.

As shown in Table 5.2, the introduction of PTCDs dramatically reduces temperature-induced delay uncertainty at the cost of increased delay and power—these increases are expected because of the small *pset* and *nset* gate voltages of the inserted PTCDs. To reduce power consumption, the PTCDs may be shut off when a path is not in use, enabling a fine-grained power gating capability (assuming the circuits can be shut off). The $PTCD_{nset+pset}$ inverter has a worst-case (i.e. 125°C, input low) leakage power 43.6% smaller than the svt inverter worst-case leakage power. When the PTCDs are shut off, the leakage reduction is 84.6%.

5.2.2 Comparison of Temperature-Insensitive Voltage Ranges

In this chapter, we have reviewed four techniques for achieving temperature insensitivity in system design—(1) the use of a technology's V_{INS} at the technology's standard threshold voltage, (2) the use of multiple V_T levels (lvt, svt, and hvt) with the V_{INS} for each device, (3) the use of adaptive body-biasing (ABB) to adjust the value of V_{INS}, and (4) the proposed programmable temperature compensation device (PTCD) approach. In Fig. 5.11, we compare the range of supply voltages that can be made temperature insensitive in each method. As shown, the single-V_T and multi-V_T methods are each fairly limited in their temperature-resilient supply voltages. The use of body biasing expands this range, yet it is still far below the nominal supply voltage; in addition, body biasing each V_T device type individually will have very large overheads, as well as requiring a triple-well process for the NMOS devices.

In contrast, the proposed approach achieves temperature insensitivity over the entire supply voltage range between 0.6 V and the nominal supply voltage, 1.0 V (assuming 'worst case insensitivity' and 1:4 P:N ratio; the dual-edge insensitivity range is 0.71—1 V), making it of greater use for dynamic voltage scaling systems.

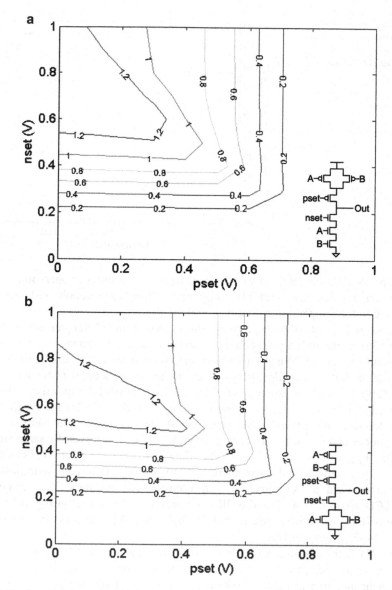

Fig. 5.12 Contour plots of (**a**) NAND gate and (**b**) NOR gate temperature dependence ($t_{125°C}$/$t_{-55°C}$) at nominal voltage

5.2.3 PTCD Integration with Other Logic Structures

The PTCD approach can also achieve temperature insensitivity in larger logic gates with more complex pull-up and pull-down networks, such as NAND and NOR gates. In Fig. 5.12 we present contour plots of the temperature dependence of

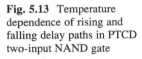

Fig. 5.13 Temperature
dependence of rising and
falling delay paths in PTCD
two-input NAND gate

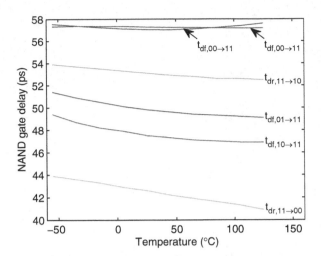

Fig. 5.13 Temperature dependence of rising and falling delay paths in PTCD two-input NAND gate

two-input NAND and NOR gates over the range of *pset* and *nset* operating points. These larger gates have multiple rising and falling edge delays, reducing the accuracy of the temperature variation compensation.

The six rising and falling edge delays for the two-input NAND gate are shown in Fig. 5.13 over the military-specified temperature range; as shown, the worst-case rising and worst-case falling edge delays are compensated to a delay variation of ~1%, although the other edge delays have a larger temperature dependence.

In Table 5.3, we compare the use of $V_{INS,P}$ (the designs labeled VINS) with the proposed $PTCD_{nset+pset}$ design style (the designs labeled PTCD) for a variety of logic structures including NAND gates, NOR gates, transmission gates, a 1-bit mirror adder and a D flip-flop; each gate in the comparison uses the same supply voltage, $V_{DD} = 710$ mV ($V_{INS,P}$). The data in Table 5.3 were collected by connecting a conventional inverter driver to each input of the gate under test; the simulations used a fixed load of four conventional inverters. The design of the PTCD gates for NAND and NOR structures are shown inset in Fig. 5.12. The transmission gate, mirror adder, and D flip-flop (DFF) designs are shown in Figs. 5.14–5.16, respectively.

The PTCD transmission gate uses a similar technique to the other gates; the *pset* and *nset* devices are placed closest to the output, with *pset* limiting the current flow through the pull-up path and *nset* limiting the current flow through the pull-down path. The mirror adder in Fig. 5.15 uses two *pset* devices and two *nset* devices, one group to achieve temperature insensitivity in the inverter *Cout* and one to generate temperature insensitivity in the inverted *Sum*. The flip-flop design uses two *pset* and *nset* devices in similar fashion, ensuring that each device chain is made temperature insensitive.

All of the PTCD logic gates in Table 5.3 use the same *pset* PTCD voltage and the same *nset* PTCD voltage to limit the integration complexity. The delay uncertainties reported in Table 5.3 are computed using the worst case rising delay and falling delay. In the larger gates, the PTCD approach is shown to reduce

Table 5.3 Logic gate comparison between V_{INS} and $PTCD_{nset+pset}$

Design	Device count	Worst case delay (ps)	Avg. power @ 27°C	P-τ product (aJ)	Min. delay uncert. (%)	Max. delay uncert. (%)
VINS$_{NAND2}$	4	53.7	485 nW	26.0	3.3	14.0
PTCD$_{NAND2}$	6	152.7	454 nW	69.5	1.6	5.2
VINS$_{NOR2}$	4	51.1	547 nW	28.0	2.4	16.8
PTCD$_{NOR2}$	6	165.1	475 nW	78.4	1.1	2.6
VINS$_{NAND3}$	6	57.2	342 nW	19.6	3.7	14.2
PTCD$_{NAND3}$	8	209.1	254 nW	53.1	2.0	7.7
VINS$_{NOR3}$	6	71.8	497 nW	35.7	9.6	20.4
PTCD$_{NOR3}$	8	236.8	284 nW	67.3	1.4	2.0
VINS$_{TG}$	2	509	28.3 nW	14.4	0.2	16.2
PTCD$_{TG}$	4	665	63.3 nW	42.1	1.2	6.2
VINS$_{MA,Co}$	24	52.3	1.98 μW	103	1.6	9.2
PTCD$_{MA,Co}$	28	145.2	1.11 μW	160	3.5	4.7
VINS$_{MA,Sum}$	24	90.4	1.98 μW	179	3.6	10.1
PTCD$_{MA,Sum}$	28	229.2	1.11 μW	254	0.5	1.4
VINS$_{DFF,C \rightarrow Q}$	8	26.5	1.56 μW	41.3	5.5	20.2
PTCD$_{DFF,C \rightarrow Q}$	12	131	1.05 μW	137	0.7	1.9
VINS$_{DFF, setup}$	8	58.6	1.56 μW	91.4	3.5	9.2
PTCD$_{DFF,setup}$	12	107	1.05 μW	112	10.1	11.2
VINS$_{DFF,hold}$	8	−5.4	1.56 μW	×	8.6	46.3
PTCD$_{DFF,hold}$	12	−49.1	1.05 μW	×	26.9	44.5

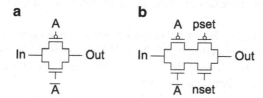

Fig. 5.14 Transmission gate design using (**a**) conventional static CMOS and (**b**) proposed PTCD methodology

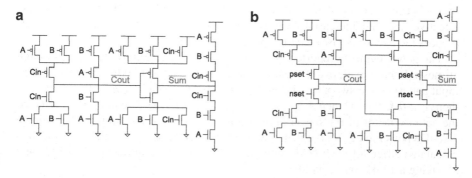

Fig. 5.15 1-bit mirror adder design using (**a**) conventional static CMOS and (**b**) proposed PTCD methodology

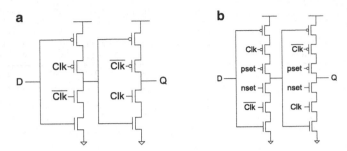

Fig. 5.16 C2MOS D flip-flop design using (**a**) conventional static CMOS and (**b**) proposed PTCD methodology

maximum delay uncertainty by up to 90%. The delay uncertainty of the V_{INS} approach increases as the number of gate inputs increases; this is because the longer device chains alter the effective threshold voltage of the PUN and PDN, changing the value of V_{INS}. The PMOS devices are insensitive to temperature at $V_{DD} = 710\,\mathrm{mV}$ while the NMOS devices are not; the device chain threshold shift further increases the PDN delay uncertainty, causing the three-input NOR gate ($VINS_{NOR3}$) uncertainty to exceed 20%. The minimum delay uncertainty in the VINS designs is achieved by the PUN, because the PMOS devices are still relatively close to their V_{INS} despite the threshold shift. In contrast, the PTCD gates, in which the PUN and PDN are each tuned to temperature insensitive biases, maintain overall lower delay uncertainty than the V_{INS} gates.

PTCD$_{NOR}$ gates are shown to maintain lower uncertainty than PTCD$_{NAND}$ gates—the maximum uncertainty of the PTCD$_{NAND3}$ is 7.7%, while that of the PTCD$_{NOR3}$ is just 2.0%. This difference is also related to the PMOS devices being more temperature resilient than the NMOS devices at $V_{DD} = 710$ mV. In NOR gates, the long device chain is in the PUN; the PMOS PTCD becomes less effective because of the PUN threshold shift, but PMOS devices are still close to their V_{INS} and remain nearly insensitive to temperature. The PTCD$_{NOR}$ PDN devices are in parallel and there is no threshold shift. In contrast, the PTCD$_{NAND}$ gates are less robust because the threshold shift is in the PDN. The NMOS PTCD becomes less effective, and the increased NMOS temperature sensitivity at $V_{DD} = 710$ mV reduces the overall NAND gate temperature resilience.

The power numbers reported in Table 5.3 were calculated by sweeping through all possible input patterns of a gate with the fastest input toggling at 1 GHz. The input patterns were swept according to an ordered truth table (e.g. $A = 0$ & $B = 0$, then $A = 0$ & $B = 1$, then $A = 1$ & $B = 0$, etc.); thus, the gate input total affected the activity factor of the power simulations, for example resulting in the average NAND3 power to be smaller than the average NAND2 power. The DFF power was calculated differently, with *Clk* having a 1 GHz period rather than pulse width, and *D* having a 1 GHz pulse width.

Table 5.3 also shows that the ratio of the number of devices in the PTCD and VINS approaches decreases as the number of gate inputs increases (layout considerations and actual area overhead are described in a later section). The PTCD inverter has twice the number of devices as the VINS inverter, a ratio of 4:2; however, in a 3-input NAND gate the ratio is reduced to 8:6.

We also examine gates with larger delays and larger power consumptions. At $V_{DD} = 710\,\mathrm{mV}$, the lower driving strengths and lack of direct connection to V_{DD} or ground make transmission gate performance the slowest of the structures evaluated. The largest unit we consider is the 1-bit mirror adder, which contains both AOI and OAI-type structures. We are able to make this design temperature insensitive using only four added PTCDs, giving it the lowest device cost of the examined units. The PTCD mirror adder (PTCD$_{\mathrm{MA}}$) has similar trade-offs to the other gate types, demonstrating the technique's scalability.

Finally, we show that sequential logic elements can also take advantage of the proposed methodology, with the PTCD version of the C2MOS DFF's clock-Q delay (PTCD$_{\mathrm{DFF,C \to Q}}$) reducing temperature sensitivity from over 22% to just 5%. DFF setup and hold time each have larger temperature sensitivity because of the changes in output slew rate as the data edge approaches the clock edge. The impact of slew rate on temperature sensitivity is provided in the discussion section later in this chapter.

Note that the designs in Table 5.3 are all for the PTCD$_{\mathrm{nset+pset}}$ methodology; the PTCD$_{\mathrm{nset}}$ methodology requires half of the additional devices and achieves smaller delay and power overheads at the cost of reduced supply voltage versatility.

5.3 Applications

In this section, we evaluate the performance of the proposed method in a set of applications that are particularly suited to take advantage of low temperature sensitivity.

5.3.1 Temperature-Insensitive Clock Tree

On-chip temperature gradients can affect the clock skew between leaf nodes, with temperature-induced skews in excess of 170 ps reported [19]. When discussing the impact of temperature on a clock tree, we must also take into account the temperature dependence of the long interconnect links between clock repeaters. The interconnect resistance R is related to temperature by

$$R(T) = R_0 \cdot [1 + \alpha_R (T - T_0)] \tag{5.5}$$

where T is the temperature, R_0 is the resistance at nominal temperature T_0, and α_R is an empirical term named the temperature coefficient of resistance. Al and Cu interconnects have similar values of α_R—0.004308 and 0.00401, respectively. Over the military-specified temperature range, Al wire resistances can change by up to 77.5% while Cu wire resistances can change by up to 72.2%. Interconnect resistance increases with increasing temperature, but this positive temperature dependence may be offset using a supply voltage where devices have a negative dependence (or zero dependence [11]) to achieve very low overall temperature sensitivity [5].

Prior work on reducing clock skew has made use of temperature-aware floorplanning [20, 21], temperature-adaptive buffers [19, 22], and the use of V_{INS} with reduced temperature-dependent level converters and frequency doublers (LCFDs) [11]. Each of these techniques are effective for reducing clock skew; however, floorplan approaches are limited by chip layout requirements and the need for complex tools, temperature-adaptive systems require the use of multiple temperature sensors in addition to the complexity of a runtime adaptive control system, and the multi-voltage framework requires the added complexity of running the clock off of a separate supply voltage network and adding LCFDs.

Alternatively, the proposed PTCD technique does not have any layout dependencies, does not require temperature sensors, control monitors, or runtime adaptivity, and uses the same clock supply voltage as the system supply voltage. Fig. 5.17 shows the clock skew with and without the proposed technique, using the balanced H-tree dimensions and sizings described in [19] and the design criteria in [11] (ensuring the clock transition is <10% of the 1 ns clock period). Note that the data in Table 5.4 is reported 'as is' from the referenced papers, and may use different topologies and technologies. The proposed method achieves clock skew reductions of up to 98.6% compared to the conventional clock tree using a worst-case thermal profile (the x-axis in Fig. 5.17 refers to the temperature difference between two leaf nodes on opposite chip corners). The addition of the PTCD devices causes an average clock tree power increase of 4% over the temperature range, although the worst-case power is reduced by 2.2%.

The skew reduction in the proposed approach is compared to prior approaches in Table 5.4. The proposed approach is shown to achieve the largest skew reduction of the compared approaches; the best prior approach reduces skew to 7.6% of the original value, while the proposed approach reduces skew to just 1.4% of the original value, an improvement of 81.6%.

5.3.2 Temperature Sensitivity Adjustment for Improved Sensor Accuracy

The proposed approach can also provide adjustable temperature sensitivity for use in sensor designs. For example, a voltage reference [23] or IR drop sensor [24] may be made temperature insensitive to improve the accuracy of a sensor reading. Reducing the impact of temperature variations on ring oscillator frequency f_{osc}

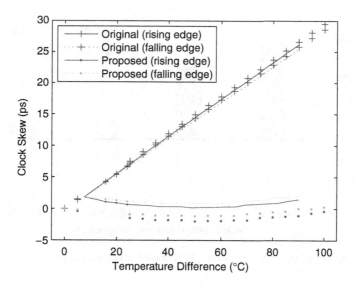

Fig. 5.17 Clock skew reduction achieved using proposed PTCD technique

Table 5.4 Clock skew reduction comparison

Technique	Skew magnitude		Skew reduction (%)
	Original (ps)	Updated (ps)	
Floorplanning [21]	2.730	650	76.2
V_{INS} + LCFD [11]	54.7	13.0	76.2
Adaptive buffers [19]	70.9	5.4	92.4
Proposed	29.5	0.4	98.6

can enable dramatic improvements in voltage variation readings. Four designs are compared in Fig. 5.18—a conventional oscillator operating at nominal voltage, a conventional oscillator operating at V_{INS} to reduce the impact of temperature variations, an oscillator using the proposed PTCD$_{nset}$ approach to minimize temperature variations (operating at V_{INS}), and an oscillator using the proposed PTCD$_{nset+pset}$ approach (operating at nominal voltage). The figure shows the impact of temperature changes (which limit the accuracy of a voltage variation measurement) over a ±10% range of voltage variations.

Temperature limits the conventional oscillator accuracy to 18.5% at V_{NOM}. Although operating at V_{INS} reduces the impact of temperature variations when voltage is fixed, V_{INS} is much lower than V_{NOM} and small changes in voltage result in a very large impact on temperature resilience, up to 21.4%. The proposed approach operating at V_{NOM} is shown to be much more effective, limiting the temperature offset to just 4.8%. Thus, the proposed approach dramatically reduces

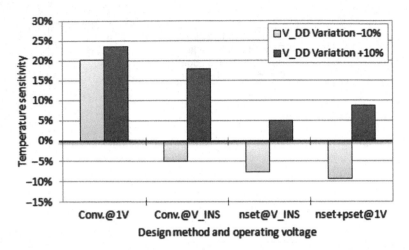

Fig. 5.18 Improved temperature resilience in the presence of voltage variations using the proposed method

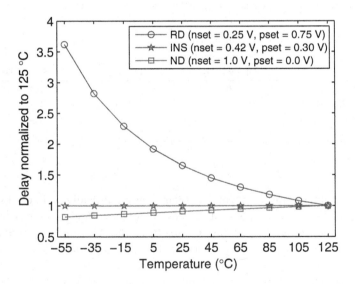

Fig. 5.19 Impact of changing *nset* and *pset* on the temperature dependence of an 11-stage ring oscillator period, normalized to 125°C

the impact of temperature variations on the ring oscillator, allowing for improved accuracy in voltage measurements.

Programmable temperature sensitivity can also be used to improve calibration and measurement capabilities of a temperature sensor; the baseline sensor reading can be taken in the temperature-insensitive state to facilitate compensation for static variations, and then the sensor's temperature sensitivity can be carefully controlled to ensure accurate operation.

The temperature dependence range achieved by the PTCD approach in an 11-stage ring oscillator (a commonly used temperature-dependent element in digital sensor designs [5, 25]) is shown in Fig. 5.19. For applications requiring strict linearity, the $-I$-T slope (normal temperature dependence) bias is within 1.1% of a linear fit. The temperature insensitive (INS) bias has a temperature sensitivity of just 0.17%, enabling excellent temperature rejection for more accurate measurement of process and/or voltage conditions. For digital set point sensors, where the output is compared to a fixed value [26], larger temperature sensitivity improves accuracy, increasing the change in value between set points; in these applications, the $+I$-T slope (reverse temperature dependence) bias would offer a 14.7× larger temperature dependence than the $-I$-T slope bias.

5.4 Discussion

The proposed approach has been shown to effectively reduce delay uncertainty and improve temperature resilience. In this section, we discuss the limitations of the approach and other design considerations.

5.4.1 Layout Area Overhead

In the PTCD$_{pset+nset}$ method, the addition of the PTCDs into the pull-up and pull-down networks results in an area overhead increase of two transistors per logic gate (although the devices are minimum sized to optimize $P*\tau$). The largest overhead occurs in the smallest gate, with a PTCD inverter having twice the transistor count of a conventional design, shown in Fig. 5.20. Although the number of transistors is

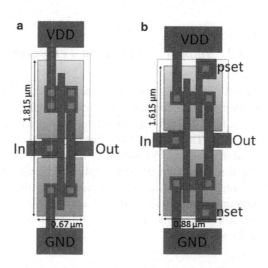

Fig. 5.20 Layout area of (**a**) conventional CMOS inverter and (**b**) PTCD inverter

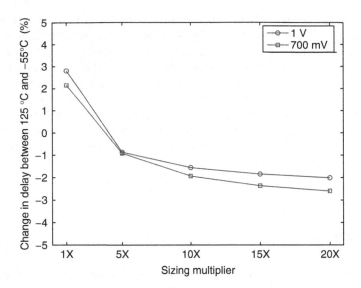

Fig. 5.21 Impact of device sizing on temperature insensitivity for the svt + nset + pset structure at 1 V and 700 mV

doubled, the actual area overhead of the PTCD approach is significantly less than 2×; the additional transistors only require an additional poly finger in both the pull-up and pull-down networks, increasing the inverter area from 1.216 to 1.421 μm^2 in the 65 nm technology—an increase of 17%. In addition to the extra devices, the PTCD method requires two additional supply voltages V_{pset} and V_{nset}, and their associated routing resources; however, this portion of the overhead is similar to that required for adaptive body-biasing methods [3, 14], which also require two additional supply voltages and associated routing resources.

5.4.2 Impact of Sizing on Temperature Insensitivity

In addition to the sizing overhead, we also examine the impact of device sizing on the temperature dependence. These simulations involved an 11-stage ring oscillator sized for a 2:1 ratio with a sizing multiplier varying between minimum size (1×) and 20×. The results, shown in Fig. 5.21, indicate that the specific device sizing used in a gate only impacts the temperature insensitivity of the delay by about ±3% for the range of sizing multipliers, showing that the proposed methodology is reasonably robust to different sizing requirements. We also show that this sizing immunity is maintained at different supply voltages, plotting curves at both 700 mV and 1 V. In contrast, conventional CMOS designs vary by ~18–19% over this range of conditions.

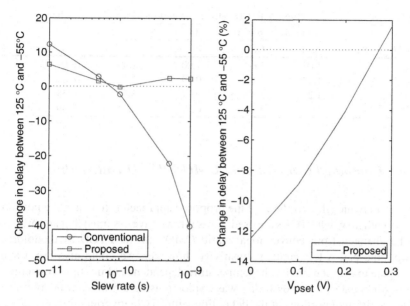

Fig. 5.22 (a) Impact of slew rate on temperature sensitivity for the conventional structure and PTCD$_{nset+pset}$ structure at 1 V, (b) impact of 1 ns slew rate on PTCD$_{nset+pset}$ for different values of V_{pset} (V_{nset} = 440 mV)

5.4.3 Impact of Slew Rate on Temperature Sensitivity

One further concern related to the device sizing is the well-known impact of changing slew rate on the temperature sensitivity [27]. For this simulation, we use a 1 V supply and drive both a conventional CMOS inverter and PTCD$_{nset+pset}$ inverter using an ideal source with varying slew rate. As shown in Fig. 5.22a, the temperature dependence of the conventional design is heavily dependent on slew rate, as expected; however, the proposed design is much less sensitive to slew rate changes, varying less than 7% from the insensitive point at a slew rate of 10 ps. To determine why the PTCD design is less sensitive to slew rate than the conventional design, we ran an additional set of simulations using a fixed slew rate of 1 ns and varying the *pset* voltage (V_{pset}).

As shown in Fig. 5.22b, the change in V_{pset} has a dramatic effect on the temperature sensitivity of gate delay, with a temperature insensitive point shown where the curve intersects y = 0. This reduction in delay uncertainty is likely caused by the current-limiting effect of the PTCDs—increasing the slew rate would normally cause the devices to go through a longer period of saturation when charging an output; however, the PTCDs limit the current when the slew rate is fast, reducing the difference between fast and slow slew rates on the pull-up and pull-down networks.

Table 5.5 Performance of inserting PTCD cells into a portion of a 5-cell delay line

# of PTCD gates	Normalized to #PTCDs = 0				Max. temp. var. (%)
	Area	Delay	Power	P*τ	
0	1.00	1.00	1.00	1.00	18.0
1	1.07	2.13	0.99	2.11	0.52
3	1.22	3.80	0.97	3.72	0.55
5	1.36	5.44	0.97	5.21	0.49

5.4.4   Combined Conventional CMOS/PTCD Datapaths

One way to reduce the overhead of the proposed approach is to adjust the portion of the data path in which PTCDs are used. Thus far, we have assumed that an entire path will be changed from conventional static CMOS to the PTCD methodology to improve the path's temperature sensitivity; however, we could also use a single PTCD gate to compensate for the temperature dependence of multiple conventional static CMOS gates. This possibility was explored using a 5-inverter delay line and replacing different portions of the delay line with PTCD inverters. In each case, we change the value of *pset* and *nset* to achieve temperature insensitivity. As shown in Table 5.5, a single PTCD gate is sufficient to achieve excellent temperature insensitivity by compensating for the sensitivity of four CMOS inverters. In addition to improving temperature resilience, this approach can significantly improve area, delay, and power-delay product ($P*\tau$) performance. The main challenge with this approach will be finding the correct number of gates to replace in each path such that only one value of *pset* and *nset* are required (of course, if a larger number of *pset* and *nset* biases are available this challenge is easily overcome).

To further reduce the overhead of the proposed approach, the methodology could be applied only to critical subsystems rather than every gate, or the PTCDs could be used as a safety mode when large temperature gradients are detected or predicted (e.g. when turning on a heater in a biological sensor lab-on-chip).

5.4.5   Impact of Wire Temperature Dependence
on PTCD Methodology

The impact of temperature variations on wire resistance can cause paths with long interconnections to have a different temperature dependence than paths with short interconnects, limiting the effectiveness of the proposed temperature insensitive technique. To study this impact, we compare the delays of PTCD datapaths with four wire lengths in Fig. 5.23—1 µm, 10 µm, 100 µm, and 1 mm—and examine some potential solutions to these large changes in sensitivity. The bold solid lines in Fig. 5.23 indicate the change in temperature dependence for unbuffered wires,

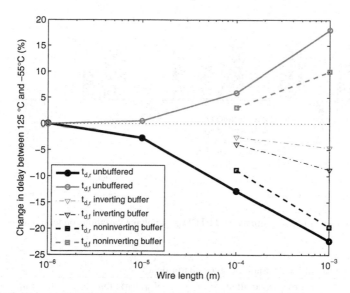

Fig. 5.23 Impact of interconnect length on achievable temperature sensitivity

shown to be temperature insensitive for the 1 μm case (using the same *pset* and *nset* voltages from earlier in this chapter), and increasing to ~20% for the unbuffered 1 mm link.

This is clearly unacceptable if we are to claim that the PTCDs enable temperature-insensitive design, so we examined the impact of buffering long interconnect links with a single two-inverter buffer (buffering is a common technique to improve communication efficiency in long interconnect), shown by the dashed lines in Fig. 5.23. While this buffer insertion offers some improvement over the unbuffered case, the buffered link still has a large temperature sensitivity; however, examination of the solid and dashed lines in Fig. 5.23 shows that the interconnect temperature dependence causes rise time to become faster and our fall time to become slower. The phenomenon behind this behavior requires further study, but we are able to exploit the behavior using inverting buffers. The use of a single inverting buffer reduces the overall datapath temperature dependence by over 60% in the 1 mm wire case, and 70% in the 100 μm case. With this inverting buffer approach, we are able to maintain near temperature insensitivity using single *pset* and *nset* values over a wide range of interconnect link lengths.

5.4.6 Variation Considerations

One of the major limitations of the proposed methodology is its susceptibility to variations. Although we have shown in Fig. 5.18 that the proposed PTCD methodology is much less susceptible to voltage variations than the conventional

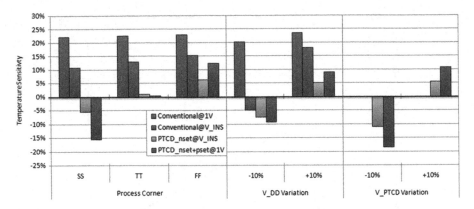

Fig. 5.24 Impact of process, supply, and PTCD gate voltage variations on temperature sensitivity

Table 5.6 Impact of PVT variations on oscillator delay

Technique	Process variation			V_{DD} variation		V_{PTCD} variation	
	SS (ps)	TT (ps)	FF (ps)	−10% (ps)	+10% (ps)	−10%	+10%
Conv. @ 1 V	11.0	8.7	6.8	11.4	8.7	N/A	N/A
Conv. @ V_{INS}	19.2	14.2	10.2	19.8	10.7	N/A	N/A
nset @ V_{INS}	42.3	29.2	19.7	36.8	30.2	27.1 ps	35.2 ps
nset + pset @ 1 V	69.0	44.8	32.1	66.1	52.8	61.6 ps	47.5 ps

design, voltage variations can be reduced to <2% with proper use of decoupling capacitors [28], limiting their impact on temperature sensitivity. For completeness, we also examine the impact of ±10% V_{pset} and V_{nset} variations in Fig. 5.24, though these voltage networks do not source large amounts of current (only gate leakage) and should have even less noise than the main voltage supply and ground lines.

Although voltage variations can be compensated by decoupling capacitors, the proposed approach is still susceptible to process variations, also shown in Fig. 5.24. Although we see that the PTCD$_{nset}$ design is much less susceptible than the other compared designs, the PTCD$_{nset+pset}$ design has a −15%/+13% range of temperature sensitivities between fast and slow process corners. Although this might appear to be the equivalent of a 28% sensitivity range, the difference between −15%/+13% and +28% is significant—a chip diagnosed to be in the SS corner can have V_{pset} and V_{nset} adjusted to compensate for the process variations after manufacture (e.g. storing the values in a ROM), limiting the temperature sensitivity to either −15% *or* +13%. In contrast, the +28% system will have 28% temperature sensitivity regardless of the process corner. Despite this consideration, the increase in temperature sensitivity is clearly unacceptable given the large overheads we are imposing, so we have developed a runtime compensation for process variations, discussed in the following section. For completeness, we have included the raw data used to generate the variation charts in Table 5.6.

5.5   Compensation for Process Variations and Aging

To maintain temperature insensitivity in the presence of process variations, a method of detecting process-induced temperature dependence (and the direction of the temperature dependence—positive or negative) is needed. If the circuit delay is found to increase with increasing temperature, we can adjust *nset* and *pset* to restore insensitivity. How to change the values of *pset* and *nset* can be determined using Fig. 5.25, which plots the $t_{125°C}/t_{-55°C} = 1$ contours for the fast (FF), typical (TT), and slow (SS) process corners. As shown, in the FF corner we need to increase the value of *pset* and decrease the value of *nset* to compensate for the change in sensitivity (before the adjustment, a system designed for TT will have increasing delay with increasing temperature); similarly, if the circuit delay is found to decrease with increasing temperature (the SS case), we can decrease the value of *pset* and increase the value of *nset*.

This adjustment could be achieved using a modified dynamic voltage and frequency scaling (DVFS) system, shown in Fig. 5.26. In a conventional DVFS system [14], a target frequency is provided to the system according to the application. A temperature subsystem is used to detect thermal emergencies; if a temperature sensor detects that the target frequency cannot be met because of thermal issues, the throttle LUT reduces the target frequency to an acceptable level. This new target frequency is then passed on to the DVFS subsystem, which uses another LUT to determine the appropriate supply voltage to meet that target frequency.

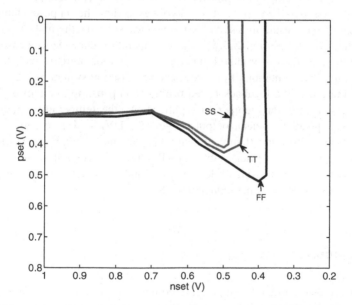

Fig. 5.25 Impact of process variations on a single gate's temperature sensitivity. Contours shown indicate $t_{125°C}/t_{-55°C} = 1$ for TT, SS, and FF corners

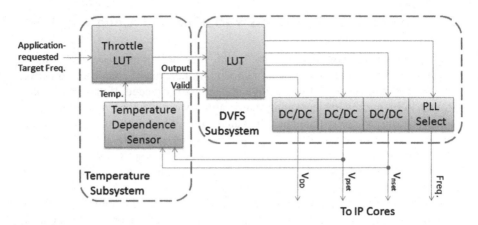

Fig. 5.26 Proposed temperature-insensitive DVFS system with thermal throttling and temperature dependence feedback for compensation of process variations and aging

To incorporate the PTCD method, two additional DC/DC converters (resulting in overhead similar to approaches such as adaptive body biasing [3, 14]) are needed to provide *pset* and *nset*. To compensate for the impact of process variations on the temperature sensitivity, a modified thermal sensor is needed that can detect the temperature dependence [29, 30] using the current values of *pset* and *nset*—achieved using a ring oscillator of PTCD-style inverters. The temperature dependence is passed to the DVFS LUT, which can adjust the values of *pset* and *nset* until the desired level of temperature insensitivity is restored. The detected temperature must still be passed to the throttle LUT to avoid exceeding the supported temperature range, which may result in thermal emergencies such as thermal runaway and decrease the overall chip reliability [31]. A temperature dependence sensor design achieving these goals was proposed in Chap. 3. The temperature dependence sensor will facilitate implementation of the process-compensation system in Fig. 5.26.

The operation of the feedback-based process compensation system is shown in Fig. 5.27 for an arbitrary temperature pattern. As the temperature changes, the temperature dependence sensor output triggers the DVFS LUT to compensate for the large positive delay uncertainty, reducing V_{nset} and increasing V_{pset}. As shown, the temperature dependence sensor enables this PTCD tuning regardless of whether the actual temperature is increasing or decreasing, reducing the temperature dependence from 8% to just over 1%, a net reduction of 87%.

5.6 Summary

The proposed programmable temperature compensation device (PTCD) methodology is a more versatile alternative to the use of a technology's temperature-insensitive supply voltage V_{INS}. PTCDs reduce delay uncertainty by up to 91%

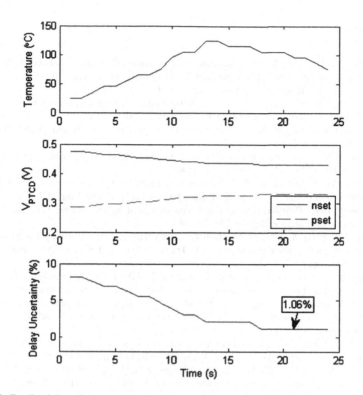

Fig. 5.27 Feedback-based process compensation results

compared to a conventional design at V_{INS} by individually compensating for temperature-induced delay variations in the pull-up and pull-down networks of each logic gate. The introduction of the PTCDs can also enable fine-grained power gating by turning off the PTCDs.

The PTCD approach achieves temperature insensitivity over a wider voltage range than the use of body-biasing and V_{INS}. We present the first demonstration of temperature insensitivity at a technology's nominal supply voltage (1 V); previous methods of achieving temperature insensitivity using V_{INS}, multi-V_T devices, and adaptive body-bias are restricted to voltages less than 85% of nominal voltage.

Specific applications of the approach include reducing delay uncertainty (clock skew is reduced by 82% compared to prior approaches) and improving sensor accuracy (temperature offsets in a voltage variation sensor are improved by 74%).

Limitations of the proposed approach include area and energy overheads and an increased susceptibility to process variations. The presented feedback system can reduce process-induced delay uncertainty by up to 87%, adaptive PTCD gate biases to the temperature insensitive state. Other limitations include the additional voltage networks required to supply PTCD voltages, as well as characterization and verification challenges if applying the PTCD method to achieve temperature insensitivity at multiple supply voltage.

References

1. Filanovsky IM, Allam A (2001) Mutual compensation of mobility and threshold voltage temperature effects with applications in CMOS circuits. IEEE Trans Circuits and Syst I: Fundamental Theory and Applications 48:876–884
2. Ku JC, Ismail Y (2007) On the scaling of temperature-dependent effects. IEEE Trans Computer-Aided Design of Integrated Circuits and Syst 26:1882–1888
3. Borkar S et al (2003) Parameter variations and impact on circuits and microarchitecture. Design Automation Conf 338–342
4. US Dept of Defense (2007) Integrated circuits (microcircuits) manufacturing, general specification, Std MIL-PRF-38535H. Washington DC
5. Wolpert D, Fu B, Ampadu P (2010) Temperature-aware delay borrowing for energy-efficient low-voltage link design. 4th ACM/IEEE Int Symp on Networks-on-Chip 107–114
6. Park C et al (1995) Reversal of temperature dependence of integrated circuits operating at very low voltages. Int Electron Devices Mtg 71–74
7. Bellaouar A, Fridi A, Elmasry MI, Itoh K (1998) Supply voltage scaling for temperature-insensitive CMOS circuit operation. IEEE Trans Circuits and Syst II: Analog and Digital Signal Processing 45:415–417
8. Lasbouygues B, Wilson R, Azemard N, Maurine P (2006) Timing analysis in presence of supply voltage and temperature variations. ACM Int Symp on Physical Design 10–16
9. Wolpert D, Ampadu P (2008) Normal and reverse temperature dependence in variation-tolerant nanoscale systems with high-k dielectrics and metal gates. 3rd ACM Int Conf on Nano-Networks 1–5
10. Kumar R, Kursun V (2007) Voltage optimization for simultaneous energy efficiency and temperature variation resilience in CMOS circuits. Microelectronics J 38:583–594
11. Tawfik SA, Kursun V (2010) Dual supply voltages and dual clock frequencies for lower clock power and suppressed temperature-gradient-induced clock skew. IEEE Trans Very Large Scale Integr (VLSI) Syst 18:347–355
12. Calimera A, Macii E, Poncino M, Bahar RI (2008) Temperature-insensitive synthesis using multi-Vt libraries. 18th ACM Great Lakes Symp on VLSI 5–10
13. Ono G, Miyazaki M, Watanabe K, Kawahara T (2005) An LSI system with locked in temperature insensitive state achieved by using body bias technique. IEEE Int Symp on Circuits and Syst 632–635
14. Tschanz J et al (2007) Adaptive frequency and biasing techniques for tolerance to dynamic temperature-voltage variations and aging. IEEE Int Solid-State Circuits Conf 292–604
15. Wolpert D, Ampadu P (2008) A low-power safety mode for variation tolerant systems-on-chip. 23rd IEEE Int Symp on Defect and Fault Tolerance in VLSI Syst 1–9
16. Long J, Memik SO (2010) Inversed temperature dependence aware clock skew scheduling for sequential circuits. Design, Automation, and Test in Europe 1657–1660
17. Kang SM, Leblebici Y (2003) CMOS Digital Integrated Circuits, 3rd ed. McGraw-Hill, NY
18. Krupka J et al (2006) Measurements of permittivity, dielectric loss tangent, and resistivity of float-zone Silicon at microwave frequencies. IEEE Trans Microwave Theory and Techniques 54:3995–4001
19. Ragheb T et al (2009) Design of thermally robust clock trees using dynamically adaptive clock buffers. IEEE Trans Circuits and Syst I: Regular Papers 56:374–383
20. Cho M, Ahmedtt S, Pan DZ (2005) TACO: temperature aware clock-tree optimization. Int Conf Computer-Aided Design 582–587
21. Yu H, Hu Y, Liu C, He L (2007) Minimal skew clock embedding considering time variant temperature gradient. Int Symp Physical Design 173–180
22. Chakraborty A et al (2008) Implementation of a thermal management unit for canceling temperature-dependent clock skew variations. Integration, the VLSI J 41:2–8
23. Liu Y, Li Z, Luo P, Zhang B (2010) A 0.52 ppm/°C high-order temperature-compensated voltage reference. Analog Integrated Circuits and Signal Processing 62:17–21

24. Abuhamdeh Z et al (2007) Separating temperature effects from ring-oscillator readings to measure true IR-drop on a chip. IEEE Int Test Conf 1–10
25. Wang YW, Li KSM (2009) Temperature-aware dynamic frequency and voltage scaling for reliability and yield enhancement. Asia South Pacific Design Automation Conf 49–54
26. Chen P, Chen TK, Wang YS, Chen CC (2009) A time-domain sub-micro Watt temperature sensor with digital set-point programming. IEEE Sensors J 9:1639–1646
27. Lasbouygues B, Wilson R, Azemard N, Maurine P (2007) Temperature- and voltage-aware timing analysis. IEEE Trans Computer-Aided Design of Integrated Cirucits and Syst 26:801–815
28. Ji G, Arabi T, Taylor G (2005) Design and validation of a power supply noise reduction technique. IEEE Trans Adv Packaging 28:445–448
29. Wolpert D, Ampadu P (2009) A sensor to detect normal or reverse temperature dependence in nanoscale CMOS circuits. 24th IEEE Int Symp Defect and Fault Tolerance 193–201
30. Wolpert D, Ampadu P (2011) A sensor system to detect positive and negative current-temperature dependences. IEEE Trans Circuits and Syst II: Express Briefs 58:235–239
31. Pedram M, Nazarian S (2006) Thermal modeling, analysis, and management in VLSI circuits: principles and methods. Proc IEEE 94:1487–1501

Chapter 6
Exploiting Temperature Dependence in Low-Swing Interconnect Links

The network-on-chip (NoC) paradigm is a promising solution to the global communication challenges of gigascale systems [1]; however, as billions of transistors continue to be added to these nanoscale systems, power dissipation continues to be a major design constraint. A great deal of research has examined power issues in NoCs [2–5]. The network infrastructure has a large cost—up to 36% of the power dissipation in each networked tile [6]. To improve NoC power consumption, common power saving techniques such as the use of reduced swing voltages have been applied to the interconnect links [7, 8].

Unfortunately, the reduction of link voltage makes these systems much more susceptible to variations. Temperature variations have a particularly severe impact on delay in low-voltage designs, shown in Fig. 6.1 for a 65 nm technology. The error bars in Fig. 6.1 are quantified in Table 6.1, where we see that the military-specified temperature range ($-55°C$ to $125°C$ [9]) can result in delay changes in excess of 200% at 0.6 V. Another important observation from Table 6.1 is that the delay change from $-55°C$ to $125°C$ is negative at 0.6 V and positive at 1.0 V—the normal and reverse temperature dependences described in Chap. 2. The temperature-insensitive voltage V_{INS} in this technology is indicated by the smallest error bars in Fig. 6.1, corresponding to 0.8 V.

The difference between the temperature dependences at high and low voltages provides an interesting opportunity for systems with reduced link swing voltages—if the link voltage is low enough to operate in the reverse temperature dependence region, a change in temperature will cause the link delay to vary in the opposite direction of the delay in the nominal voltage router. These opposing delay variations make room for innovative new approaches to lessen the impact of temperature variations and improve system reliability and performance.

In this chapter, we propose a temperature-aware delay borrowing method that averages the impact of temperature variation on the link and the transceiver. When the links are operated below V_{INS}, the average of the reverse temperature dependence in the link and the normal temperature dependence in the transceiver reduces the impact of temperature variation on the communication system as a whole.

D. Wolpert and P. Ampadu, *Managing Temperature Effects in Nanoscale Adaptive Systems*, DOI 10.1007/978-1-4614-0748-5_6,
© Springer Science+Business Media, LLC 2012

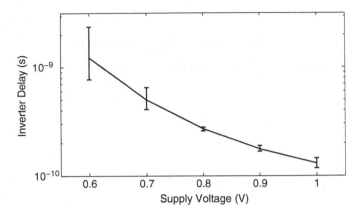

Fig. 6.1 Impact of temperature on a commercial 65 nm technology

Table 6.1 Temperature-induced delay change in a 65 nm technology

Compared temperatures	Link voltage				
	0.6 V	0.7 V	0.8 V	0.9 V	1.0 V
−55°C → 125°C	−210.0%	−59.3%	−8.2%	11.4%	19.3%
25°C → 125°C	−59.6%	−22.2%	−4.5%	4.5%	8.9%
45°C → 125°C	−42.2%	−16.5%	−3.5%	3.3%	6.8%
65°C → 125°C	−28.4%	−11.5%	−2.7%	2.23%	4.9%

6.1 Related Work

One of the simplest and most effective approaches for reducing interconnect power consumption is to reduce the supply voltage [7]. Reduced swing approaches vary widely in complexity, from a simple low-swing driver and level converter [7] to advanced signaling methods such as low-voltage differential signaling (LVDS) [10], pulsed-bus signaling [11, 12], and the use of a high-voltage boost to improve low-voltage transition delays [2, 5, 13]. Adaptive techniques such as dynamic voltage and frequency scaling are also effective [3, 14], although they require additional overhead systems. While each of these techniques reduces power consumption, low voltage links have increased susceptibility to process, voltage, and temperature (PVT) variations.

To tolerate variations, worst-case designs use PVT corner analysis to ensure that systems function properly under the majority of operating conditions. In older technologies, a single temperature corner was sufficient to determine worst-case requirements, but the need to include the reversal of temperature variation has recently led to the temperature corner being split into two separate corners [15].

Aside from the worst case corners, a large amount of research has been performed on temperature modeling to predict thermal issues at design time and examine ways of avoiding potential problems [16–19]. These temperature models have resulted in a number of design time temperature-aware techniques such as floorplanning [18], routing [20], and coding [21]. The models have also facilitated runtime techniques such as temperature-aware scheduling with traffic throttling [14, 22], and voltage/frequency throttling [23–25]. These designs do not take into account the changes in the temperature dependence at different voltages.

A number of methods have been proposed to create temperature-insensitive designs, either taking advantage of the temperature-insensitive voltage [26] or adjusting the threshold voltage to achieve temperature insensitivity [27]. Additional approaches to achieve temperature-insensitivity include the use of multiple threshold designs to balance the dependences of high-V_T and low-V_T logic cells [28]. These approaches are restricted to a select range of voltages, limiting their potential energy improvements. The approach proposed in this work takes advantage of low swing voltages *and* temperature-aware system design to improve system energy while limiting the impact of temperature variations.

6.2   Temperature-Aware Low Voltage Link Design

Temperature variation is a particularly important design consideration for low-voltage systems. One way to address these variations is the use of temperature-aware systems, which detect temperature-induced delay changes and adjust the supply voltage to maintain a target frequency [23, 29], as shown in Fig. 6.2a. When higher voltages are needed to maintain the frequency, the energy efficiency of the system is reduced.

In this chapter, we propose an alternative method of maintaining a target frequency that improves power consumption and reduces the impact of temperature variation on link delay. To achieve these improvements, we exploit the different temperature dependences in the low-voltage link and nominal voltage transceivers. As temperature decreases, link latency increases and transceiver latency decreases. To offset this increase in link latency, we adaptively borrow the additional temperature-induced slack in the receiver buffers.

To implement the proposed method, we first determine the number of stages in the transmitter and/or receiver that will be available to borrow delay slack. Then, we select the lowest voltage that can meet the target frequency at the 'slowest' temperature (e.g., −55°C for the reverse temperature dependence region) with this pre-determined timing slack; this voltage must also meet the timing requirement with no borrowed slack at the 'fastest' temperature (e.g., 125°C for the reverse temperature dependence region).

The temperature range (−55°C to 125°C) is separated into a set of temperature regions depending on the desired granularity. In Fig. 6.2b, we use four regions each spanning a 45°C range. With these regions defined, we create a look-up table that properly selects the transmitter and/or receiver buffer clock phases depending on the operating temperature. This timing borrowing greatly reduces the impact of temperature variation on system delay, enabling the proposed system to maintain a single low link voltage over the entire range of temperatures, as shown in Fig. 6.2b.

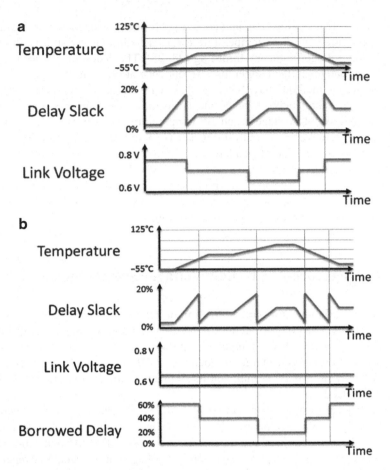

Fig. 6.2 Comparison of (**a**) conventional temperature-aware adaptive-voltage system and (**b**) proposed temperature-aware delay borrowing system

Figure 6.2 shows the timing slack in the link, which decreases with decreasing temperature until some minimum slack point is reached. At this point, in the conventional approach the link voltage is raised, which increases the slack (and consumes additional power); in the proposed approach, the decrease in temperature increases the available delay slack in the transceiver, allowing additional delay to be borrowed by the link (shown in Fig. 6.2b). By maintaining a lower voltage, the proposed system can achieve large energy savings over conventional designs.

Figure 6.3 shows the temperature-aware delay borrowing system. The output buffer of the transmitter drives a low-voltage link, with level shifters shown on the receiver end that restore the signal swing for the nominal voltage receiver buffer. On the link, a temperature sensor provides regular status updates to the delay-borrowing units in the transmitter and receiver. Temperature changes at a rate of approximately 1 µs/°C [30], so the sensor polling rate is on the order of 1 MHz,

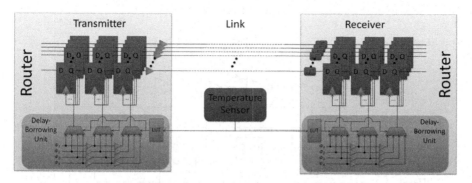

Fig. 6.3 Diagram of temperature-adaptive delay borrowing system

depending on the desired accuracy. In the delay-borrowing units, the temperature sensor output is converted into a clock phase select signal by the LUTs, and the appropriate phase is applied to the buffers. For example, at nominal voltage the transceiver buffer Clk→Q delay varies by 10.6% over the −55°C to 125°C temperature range.

As temperature decreases, the link delay increases and the buffer delay decreases; the nominal-voltage buffer will have up to 10.6% delay slack that can be borrowed by the link at low temperatures. If the delay borrowing unit is designed to trade delay with two buffers, the total amount of delay slack becomes 21.2%, etc. As shown in the system diagram in Fig. 6.3, additional buffers (or nominal voltage logic stages) may be added to the delay borrowing units by adding a small delay element to further offset the clock phase of subsequent stages.

The design of the temperature sensor and LUT are shown in Figs. 6.4 and 6.5. The temperature sensor in Fig. 6.4 is a simplified form of the one presented in Chap. 3, consisting of a temperature-dependent ring oscillator gated by an enable signal and a pulse counter to convert the oscillator frequency into a digital output [31]. For each sensor reading, the enable signal is applied for a fixed period depending on the desired sensor resolution (as the enable period increases, the impact of a small change in temperature causes a larger change in the pulse count). When the enable signal is asserted, the isolation circuitry at the bottom of Fig. 6.4 separates the temperature sensor output from the rest of the system to prevent unnecessary toggling. After the enable signal is unasserted, the pulse counter passes a digital readout of the temperature to the LUT.

The LUT in Fig. 6.5 sets the vector *Out* to '11_2' when temperatures lower than −10°C are detected, '10_2' when temperatures lower than 35°C are detected, and so on as shown in Fig. 6.6, which is the simulated response of the temperature sensor and look-up table in a 65 nm technology. As shown, both the temperature sensor and look-up table are very small circuits, with very little cost in terms of complexity or overhead energy. For a 200 ns enable period, the worst-case energy per sensor reading is 14.5 pJ, which provides a worst-case temperature resolution of 5°C.

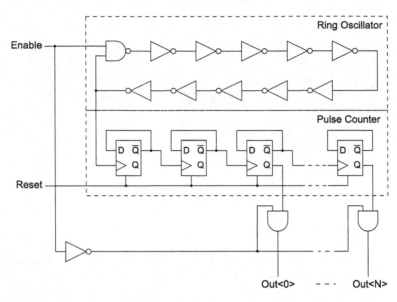

Fig. 6.4 Temperature sensor implementation for temperature-adaptive delay-borrowing system

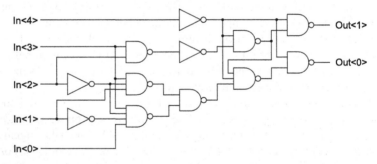

Fig. 6.5 Look-up table to select delayed clock phase based on temperature input

An example timing diagram of the system operation is shown in Fig. 6.7. The initial temperature is set to 28°C, resulting in a temperature sensor output *Sensor_Output*='51' and *LUT_Value* of '10_2.' A change in temperature is detected after some number of clock cycles (depending on the sensor sampling rate and enable period), resulting in a change in *Sensor_Output*. After the enable signal in the temperature sensor is unasserted, the sensor output is passed on to the LUT, which determines if the system has transitioned to a new temperature region and adjusts the delay-borrowing unit accordingly. In Fig. 6.7, the LUT value transitions from the '10_2' region to the '01_2' region, applying a new clock phase to the flip-flop. The system then continues polling for new changes in temperature.

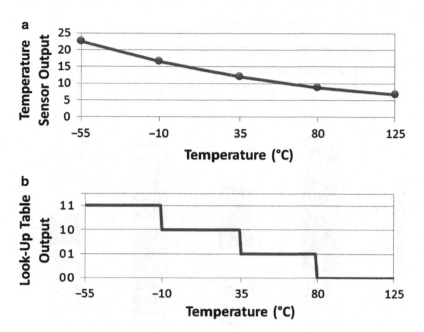

Fig. 6.6 Operation of (a) temperature sensor and (b) LUT

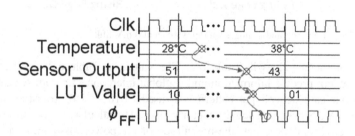

Fig. 6.7 Timing diagram of temperature-aware delay borrowing

6.3 Results

In this section we demonstrate the functionality of the proposed temperature-aware delay borrowing system.

6.3.1 Simulation Setup

Each of the simulations presented in this paper was generated using a 65 nm technology with low power svt devices. The link was simulated using a distributed RC link model of three parallel wires, with 1 mm segments and the global wire and

Table 6.2   Global wire parameters

Parameter	Value
R (Ω/mm) @ 27°C	75
Substrate capacitance, C_g (fF/mm)	75
Coupling capacitance, C_C (fF/mm)	85

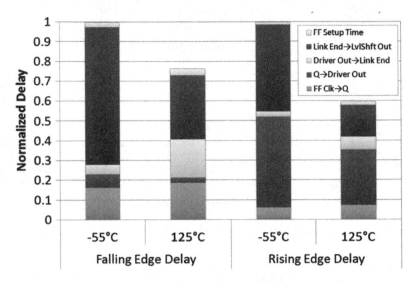

Fig. 6.8   Piecewise rising and falling edge delay for a 600 mV link

coupling parameter values in Table 6.2. The values in Table 6.2 were generated using the 65 nm process information and a global interconnect parameter calculator [32]. A worst-case delay input pattern was used ('010'→'101' transitions) for all simulations. A 0.5 switching activity factor was used to calculate the power consumption. The activity factor affects the magnitude of the power dissipation in both the proposed and conventional methods; however, the percentage power improvements achieved by the proposed method are independent of the switching factor.

6.3.2   System Characterization

Before evaluating the benefits of the proposed delay borrowing scheme, we study the impact of changing temperature eon the rise and fall time of a low voltage link. In Fig. 6.8, we separate the rising and falling edge delays into five segments—the Clk→Q delay of the transmitter flip-flop, the driver latency (Q→Driver Out), the interconnect latency (Driver Out→Link End), the level shifter latency (Link End→LvlShiftOut), and the setup time of the receiver flip-flop. The rising and falling edge delays are normalized to the −55°C condition to show which components increase or decrease in delay (as well as the change in total path

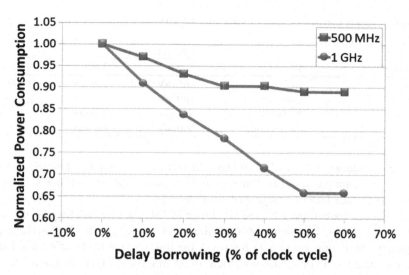

Fig. 6.9 Impact of delay borrowing on power consumption

delay) over the entire temperature range. The output of the transmitter flip-flop has a full swing between 0 V and $V_{DD} = 1$ V, while the low-voltage driver is an inverting buffer with supply voltage V_{Low} (in Fig. 6.8, $V_{Low} = 600$ mV).

As expected, the nominal voltage flip-flop delay *increases* as temperature increases, while the low-voltage driver delay *decreases* as temperature increases. The interconnect delay also increases with increasing temperature, as described in Chap. 2. The level shifter used in these simulations is the conventional level shifter described in [7]; its delay decreases with increasing temperature. Finally, the setup time requirement for the receiver flip-flop increases with increasing temperature. It is important to be aware of the difference in rising and falling delays to avoid race conditions when using delay borrowing.

6.3.3   Comparison with Conventional Low-Swing Link

As described in Fig. 6.2, delay borrowing enables us to achieve a target frequency using a lower link voltage than would otherwise be possible. The amount of voltage improvement depends on the percentage of the clock cycle that is borrowed, which in turn depends on the number of transceiver buffers used to retime the link. The power consumption in Fig. 6.9 is normalized to the 0% delay borrowing case. As shown, the improvement in power achieved by delay borrowing depends on both the target frequency and the borrowed portion of the clock cycle. At a 1 GHz frequency, borrowing 60% of the cycle period results in a power reduction of 34% in a 3 mm link. At 500 MHz, the power reduction is considerably less. The benefits are reduced at low frequencies because the target frequencies can be met at lower

Table 6.3 Supply voltages satisfying 1 GHz and 500 MHz frequency requirements in proposed and conventional methods

Target frequency	Method	Link length (mm)		
		1 mm	2 mm	3 mm
1 GHz	Proposed	605 mV	645 mV	675 mV
	Conventional	660 mV	720 mV	780 mV
500 MHz	Proposed	565 mV	595 mV	615 mV
	Conventional	580 mV	615 mV	640 mV

voltages, leading to a smaller voltage difference between the proposed delay borrowing technique and the conventional technique (at low voltages, a small change in voltage causes a large change in delay).

The voltage reductions achieved are shown in Table 6.3 for link lengths of 1 mm, 2 mm, and 3 mm, borrowing 60% of the clock cycle. At the 500 MHz target frequency, delay borrowing only improves the link voltage by 15 mV for a 1 mm link, compared to a voltage improvement of 55 mV for a 1 GHz frequency target. The results in this chapter were generated by setting a target frequency and lowering the voltage in 5 mV increments until the receiver flip-flop correctly latched the data at $-55°C$ (the worst-case temperature in the reverse temperature dependence region). Figure 6.10 compares the power consumption of the proposed delay borrowing technique and conventional low-voltage link at the frequencies and link lengths from Table 6.3. Reported power consumptions are for a 64-bit link with 0.5 activity factor at the voltages indicated in Table 6.3. As shown, the improvement in power consumption increases for longer link lengths and faster frequencies; these require larger voltages to meet the target frequency, resulting in larger voltage differences between the methods and larger improvements in power performance.

Figure 6.11 shows the improvement in power consumption over the entire temperature range for a 3 mm link achieving a 1 GHz frequency target. The link voltages are optimized for the four temperature regions in Fig. 6.6b; in the proposed approach, 60% of the clock cycle delay is borrowed when $-55°C \leq T < -10°C$, 40% delay is borrowed when $-10°C \leq T < 35°C$, 20% delay is borrowed when $35°C \leq T < 80°C$, and no delay is borrowed when $80°C \leq T < 125°C$. The maximum power savings (40%) occurs at 5°C.

6.3.4 Comparison with Temperature-Insensitive Operation

Another common use of low-voltage design is to take advantage of the temperature-insensitive voltage (V_{INS}), at which a change in temperature has very little impact of system power and delay. This design decision is an excellent choice for improving system reliability [26]; however, operating at V_{INS} imposes specific constraints on both power consumption and latency. In the commercial 65 nm technology used in

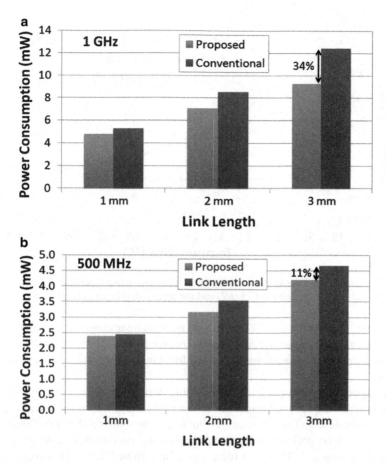

Fig. 6.10 Maximum power savings of the proposed method over conventional low-swing approach vs. link length at (**a**) 1 GHz and (**b**) 500 MHz

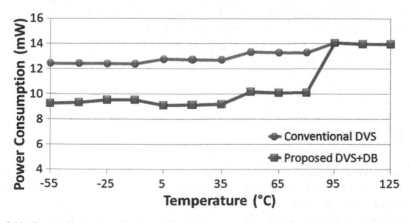

Fig. 6.11 Power dissipation of proposed and conventional methods over the temperature range for a 3 mm link designed to meet a 1 GHz target frequency

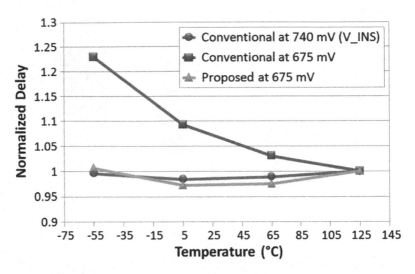

Fig. 6.12 Comparison of the delay variation of a conventional system operating at 740 mV (V_{INS}), a conventional system operating at 675 mV, and the proposed system operating at 675 mV

this chapter, V_{INS} occurs at ~740 mV; thus, systems operating at V_{INS} are incapable of taking advantage of the high speed performance of normal voltage designs, while also being unable to take advantage of the low power performance of low-voltage designs.

One aspect of the temperature-aware delay borrowing system that we have not yet discussed is the reduced impact of temperature variation on the overall system (including the link and any transceiver buffers being borrowed from). Figure 6.12 compares the delay performance of the proposed approach operating at 675 mV (the voltage achieving a 1 GHz target for a 3 mm link, from Table 6.3), a conventional link operating at 675 mV, and a conventional link operating at V_{INS}.

The delays in Fig. 6.12 are normalized to the values at 125°C. The system operating at V_{INS} has a delay variation of 1.6% over the entire temperature range, while the proposed method has a 3.4% delay variation over the entire range. Without the use of delay borrowing, the temperature-induced delay change between −55°C and 125°C exceeds 23%. Thus, the delay borrowing method improved robustness to temperature variations by over 85%.

6.3.5 Area Overhead

Unit areas were calculated by summing the standard cell widths for each standard gate and using the sum of the gate width and the nwell design rule for the link driver and level shifter, all from a 65 nm process. The total area of the temperature sensor in Fig. 6.4 is 122.4 μm^2, the total area for the LUT in Fig. 6.5 is 61.4 μm^2, and the area for each delay borrowing unit in Fig. 6.3 (including the LUT, three 4:1 multiplexors, and eight buffers) is 131.7 μm^2. Each link wire is assumed to have three transmitter flip-flops and three receiver flip-flops; combined with the driver and level shifter (ignoring the global wire area), each wire transceiver has a total area of 247.4 μm^2. Thus, for a 16-bit link, the area overhead of the proposed approach shown in Fig. 6.3 (with two delay borrowing units and one temperature sensor) is 9.7%. For a 64-bit link, the area overhead is reduced to 2.4%. Thus, multiple sensors may be integrated on each link at a relatively low overhead cost.

6.4   Discussion

In this section, we address a number of related points including integrating multiple temperature sensors, integrating the proposed design with error control coding (a commonly used reliability method for protecting interconnect links), and the temperature-dependent behavior of the low-voltage level converters (which can have devices functioning in both the normal and reverse temperature dependences within the same circuit).

6.4.1   Integration of Multiple Temperature Sensors

For simplicity, we have assumed that the link temperature profile is uniform, although this is clearly not always the case [16, 17]. For non-uniform temperature profiles, the proposed system may still be used, although some additional overhead is required. Depending on the expected temperature profiles, multiple temperature sensors may be placed along the link or at the link endpoints. To integrate these into the proposed system, the look-up tables can be adjusted to select a weighted average of the sensor inputs, and the clock phases must be adjusted to ensure that the worst-case temperature requirements are met. Exponentially-distributed thermal profiles have been shown to induce up to 7% error in circuit delay calculations compared to using an average link temperature, while a poorly located temperature sensor can result in delay errors in excess of 20% depending upon the gradient of a local hotspot [16].

6.4.2   Integration with Error Control Coding

One other important assumption that we have made is that the link is reliable and not susceptible to errors; of course, this is rarely the case. There are a number of solutions to improve the reliability of on-chip interconnect links [33–35], such as the use of error control coding. Error control encoders and decoders slightly complicate the

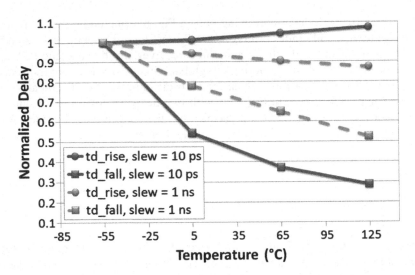

Fig. 6.13 Level shifter delays with $V_{Low} = 0.6$ V and $V_{High} = 1.0$ V for two input slew rates

delay-borrowing scheme, although the system can still function as described with a minor modification. Rather than progressively borrowing the delay from the buffers, the buffer clock phases must be adjusted to skip the stage of the encoder and decoder, which likely would not have as much slack as the neighboring empty buffer stages.

6.4.3   Level Shifter Design for Systems with Multiple Temperature Dependences

Level shifters are a particularly interesting topic when discussing multiple temperature dependences, as they operate at both V_{High} and V_{Low}. The internal inverter uses a low-voltage supply, increasing in speed as temperature increases, while the full-swing cross-coupled inverters operate at nominal voltage, decreasing in speed as temperature increases.

In Fig. 6.13, we examine the rising and falling edge delays (td_rise and td_fall, respectively) for a level shifter with $V_{Low} = 0.6$ V and $V_{High} = 1$ V driven by an input with two different slew rates. As shown, for a fast slew rate (10 ps) the rising delay increases with increasing temperature, while the falling delay decreases with increasing temperature. The different temperature dependence of the rise and fall times is related to the different impact of temperature on NMOS and PMOS devices [36]; NMOS devices operate in the reverse temperature dependence region at high voltages than PMOS devices. This causes the falling edge delay to decrease with increasing temperature, while the rising edge delay increases with increasing temperature. When the slew rate is increased to 1 ns, both the rising and falling delays decrease with increasing temperature (this matches the level shifter response shown in Fig. 6.8);

thus, the temperature response of the level shifter depends on both the level shifter P/N ratios and the input slew rate. These dependences (as well as other issues in multiple dependence level shifters) have been examined in detail in [37].

6.5 Summary

In this chapter we have presented a temperature-aware delay borrowing system for low-voltage interconnect links. The system is shown to achieve improvements in power consumption of 40% as well as an 85% reduction in susceptibility to temperature variation. The combination of improved power performance and reliability come at a very low overhead cost, less than 1% energy overhead for a 64-bit link. In addition, the reliability improvements make the proposed method a viable replacement for use of the temperature insensitive voltage in low-swing link designs, allowing delay variations as low as 3.4% while enabling the use of a wider supply voltage range to improve performance.

References

1. Dally WJ, Towles B (2001) Route packets, not wires: on-chip interconnect networks. Design Automation Conf 684–689
2. Raghunathan V, Srivastava MB, Gupta RK (2003) A survey of techniques for energy efficient on-chip communications. Design Automation Conf 900–905
3. Worm F, Ienne P, Thiran P, DeMicheli G (2005) A robust self-calibrating transmission scheme for on-chip networks. IEEE Trans Very Large Scale Integr (VLSI) Syst 13:126–139
4. Banerjee K, Mehrotra A (2002) A power-optimal repeater insertion methodology for global interconnects in nanometer design. IEEE Trans Electron Devices 49:2001–2007
5. Kaul H, Sylvester D (2005) A novel buffer circuit for energy efficient signaling in dual-VDD systems. 15th ACM Great Lakes Symp on VLSI 462–467
6. Wang H, Peh LS, Malik S (2003) Power-driven design of router microarchitectures in on-chip networks. 36th IEEE/ACM Int Symp on Microarchitecture 105–116
7. Zhang H, George V, Rabaey JM (2000) Low-swing on-chip signaling techniques: effectiveness and robustness. IEEE Trans Very Large Scale Integr (VLSI) Syst 8:264–272
8. Venkatraman V, Anders M, Kaul H, Burleson W, Krishnamurthy R (2006) A low-swing signaling circuit technique for 65 nm on-chip interconnects. IEEE Int SoC Conf 289–292
9. US Dept of Defense (2007) Integrated circuits (microcircuits) manufacturing, general specification, Std MIL-PRF-38535H. Washington DC
10. Chen M, Silva-Martinez J, Nix M, Robinson ME (2005) Low-voltage low-power LVDS drivers. IEEE J Solid-State Circ 40:472–479
11. Wang P, Pei G, Kan ECC (2004) Pulsed wave interconnect. IEEE Trans Very Large Scale Integr (VLSI) Syst 12:453–463
12. Deogun HS, Sanger R, Sylvester D, Brown R, Nowka K (2006) A dual-VDD boosted pulsed bus technique for low power and low leakage operation. IEEE Symp Low Power Electronics and Design 73–78
13. Fu B, Wolpert D, Ampadu P (2009) Lookahead-based adaptive voltage scheme for energy-efficient on-chip interconnect links. 3rd ACM/IEEE Int Symp on Networks-on-Chip 54–64

14. Shang L, Peh LS, Jha NK (2003) Dynamic voltage scaling with links for power optimization of interconnect networks. 9th Int Symp on High-Performance Computer Architecture 91–102
15. Dasdan A, Hom I (2006) Handling inverted temperature dependence in static timing analysis. ACM Tran Design Automation of Electronic Syst 11:306–324
16. Ajami A, Banerjee K, Pedram M (2005) Modeling and analysis of nonuniform substrate temperature effects on global ULSI interconnects. IEEE Trans Computer-Aided Design of Integrated Circuits 24:849–861
17. Xu S, Benito I, Burleson W (2007) Thermal impacts on NoC interconnects. IEEE Int Symp on Networks-on-Chip 220
18. Tsai JL et al (2006) Temperature-aware placement for SoCs. Proc IEEE 94:1502–1518
19. Pedram M, Nazarian S (2006) Thermal modeling, analysis, and management in VLSI circuits: principles and methods. Proc IEEE 94:1487–1501
20. Zhang T, Zhan Y, Sapatnekar SS (2006) Temperature-aware routing in 3D ICs. Asia and South Pacific Design Automation Conf 309–314
21. Wang F, Xie Y, Vijaykrishnan N, Irwin MJ (2006) On-chip bus thermal analysis and optimization. Design, Automation, and Test in Europe 850–855
22. Xie Y, Hung WL (2006) Temperature-aware task allocation and scheduling for embedded multiprocessor systems-on-chip (MPSoC) design. J VLSI Signal Processing 45:177–189
23. Tschanz J et al (2007) Adaptive frequency and biasing techniques for tolerance to dynamic temperature-voltage variations and aging. IEEE Int Solid-State Circuits Conf 292–604
24. Wolpert D, Ampadu P (2008) Adaptive delay correction for runtime variation in dynamic voltage scaling systems. J Circuits, Systems and Computers 17:1111–1128
25. Herbert S, Marculescu D (2008) Variability-aware frequency scaling in multi-clock processors. In: Adaptive Techniques for Dynamic Processor Optimization. Springer, US
26. Bellaouar A, Fridi A, Elmasry MI, Itoh K (1998) Supply voltage scaling for temperature-insensitive CMOS circuit operation. IEEE Trans Circuits and Syst II: Analog and Digital Signal Processing 45:415–417
27. Kumar R, Kursun V (2006) Supply and threshold optimization for temperature insensitive circuit performance: a comparison. IEEE Int SoC Conf 89–90
28. Calimera A, Macii E, Poncino M, Bahar RI (2008) Temperature-insensitive synthesis using multi-Vt libraries. 18th ACM Great Lakes Symp on VLSI 5–10
29. Elgebaly M, Sachdev M (2007) Variation-aware adaptive voltage scaling system. IEEE Trans Very Large Scale Integr (VLSI) Syst 15:560–571
30. Bernstein K et al (2006) High-performance CMOS variability in the 65-nm regime and beyond. IBM J Res and Dev 50:433–449
31. Wang YW, Li KSM (2009) Temperature-aware dynamic frequency and voltage scaling for reliability and yield enhancement. Asia and South Pacific Design Automation Conf 49–54
32. Arizona State University (2007) Predictive technology model [Online] http://www.eas.asu.edu/~ptm
33. Bertozzi D, Benini L, DeMicheli G (2005) Error control schemes for on-chip communication links: the energy-reliability tradeoff. IEEE Trans Computer-Aided Design of Integrated Circuits and Syst 24:818–831
34. Fu B, Ampadu P (2009) On Hamming product codes with type-II hybrid ARQ for on-chip interconnects. IEEE Trans Circuits Syst I: Regular Papers 56:2042–2054
35. Dumitras T, Marculescu R (2003) On-chip stochastic communication. Design, Automation and Test in Europe 10790–10795
36. Wolpert D, Ampadu P (2008) Normal and reverse temperature dependence in variation-tolerant nanoscale systems with high-k dielectrics and metal gates. 3rd ACM Int Conf on Nano-Networks 1–5
37. Wolpert D, Ampadu P (2010) Level shifter speed, power, and reliability trade-offs across normal and reverse temperature dependences. 53rd IEEE Int Midwest Symp on Circuits and Syst 1254–1257

Chapter 7
Avoiding Temperature-Induced Errors in On-Chip Interconnects

On-chip interconnect links become more susceptible to faults as technology scales. While most faults are temporary, about 20% of all errors are caused b y permanent or intermittent (lasting several cycles) faults [1]. These faults can occur because of manufacturing defects or run-time variations, such as multi-cycle delay failures during extended high temperature conditions or permanent faults caused by thermal runaway. Error control coding (ECC) techniques are commonly used to address reliability issues in on-chip interconnects [2–8], but these techniques generally target transient errors rather than long-duration errors. A single long-duration fault can drastically reduce or even eliminate the correction capabilities of most commonly used codes.

To maintain coding strength in the presence of permanent errors, spare wires can be used to replace permanently erroneous wires. The introduction of spare wires requires reconfiguration control logic for bypassing erroneous wires, as well as a protocol for synchronizing information between receiver and transmitter.

In this chapter, we propose an in-line test (ILT) system [9] that uses spare wires to bypass long-duration errors without interrupting the data flow. The system cycles through each adjacent pair of wires testing for opens and shorts. These tests can be run periodically to ensure that each link's ECC capability is not being crippled by permanent errors. By testing every wire in the link, the ILT method also recovers wire resources from intermittent errors that were incorrectly flagged as permanent.

ECC research in on-chip networks typically only considers transient errors; protection against permanent or intermittent errors is rarely discussed. Some methods of protecting against transient errors can also be used to protect against permanent and intermittent errors; unfortunately, this can severely limit a code's ability to protect against transient errors.

D. Wolpert and P. Ampadu, *Managing Temperature Effects in Nanoscale Adaptive Systems*, DOI 10.1007/978-1-4614-0748-5_7,
© Springer Science+Business Media, LLC 2012

7.1   The Need for Error Protection in On-Chip Interconnects

Errors in interconnect links may be caused by PVT variation, aging, particle strikes, crosstalk, or other noise sources. Some applications can tolerate errors without any major impact on functionality, for example audio or video streaming where occasional artifacts are an inconvenience rather than a catastrophic failure. Other applications have no error tolerance, such as financial databases where a single bit of corruption could result in the loss of billions of dollars.

Before describing the fundamentals of error control, we characterize the types of errors affecting our systems. There are three classes of errors: transient errors, lasting for only one or a couple of cycles, intermittent errors, lasting for many cycles, and permanent errors, lasting for the remaining lifetime of the system.

7.1.1   Transient Errors

Transient errors are somewhat rare (probabilities on the order of 10^{-9}–10^{-12} errors/ bit are commonly used [10]), and can be caused for a variety of reasons, including:

- A combination of PVT variation, crosstalk and other noise sources which is larger than the allotted frequency guardband,
- Radiation-induced single-event upsets (SEUs) from the package, called alpha particle strikes, with emission rates of 1–100 particles/cm-kh [11],
- Radiation-induced multi-bit upsets (MBUs, which we will also refer to as burst errors) from the package, which occur in up to 0.004% of all alpha particle strikes (a function of particle energy and Q_{crit} in the cells) [11],
- SEUs and MBUs from cosmic rays, which have been shown to depend on altitude (i.e., how much protection our atmosphere can provide); at sea level cosmic rays occur at a rate of approximately 10 particles/cm/h, though they are particularly problematic for avionics and space applications, with error rates of up to 1,000 particles/cm-h reported [12].

The error rates mentioned are dependent on Q_{crit}, the amount of charge required to cause an error on a node. System soft error rates have been shown to increase by two orders of magnitude from 180 nm technology to 45 nm technology because of the reduction in Q_{crit} [12].

7.1.2   Intermittent and Permanent Errors

An intermittent error is loosely defined as an error lasting for more than one or two clock cycles that is not permanent. While a transient error can be handled by a variety of error control methods, an intermittent error limits the number of correction options; for example, retransmission of information is not useful if the retransmitted information is also corrupted.

Intermittent errors can be caused by long-term effects, for example a wire which was very close to failing the initial yield test may slow because of electromigration, causing it to fail whenever temperatures approach the maximum rated value. This error could potentially last for hundreds of cycles, but is not a permanent error because when the temperature profile is altered it could still function correctly (if the wire cannot function in any condition, it is considered to be a permanent error).

Determining whether a long-term error is intermittent or permanent will be shown to cause some difficulty later in this chapter; the system should be reconfigured in the presence of a permanent error, but we must also try to avoid wasting reconfiguration resources if the error is intermittent. This decision is related to the amount of wasted energy spent correcting an error each cycle, the amount of energy required to reconfigure the system to avoid the error, and whether or not the reconfigured resources can be restored to functionality if an error turns out to be intermittent.

7.2   Error Control Coding Fundamentals

In this section we provide a background on the error control techniques used in this chapter. For a more complete discussion of the follow concepts, we recommend the extensive review on error control coding by Lin and Costello [13].

7.2.1   Triple Modular Redundancy

One of the simplest error control codes is a repetition code, in which each message bit is replicated multiple times (for example, '1,0,1' might be coded as '111,000,111'). These messages are decoded using a majority voting circuit in the receiver, and can detect $d_{min}-1$ errors or correct $\lfloor (d_{min}-1)/2 \rfloor$ errors, where d_{min} is the minimum Hamming distance (the minimum number of bit positions which are different between any two code words). In repetition codes, d_{min} is the number of times each bit is replicated. One popular example of the repetition code is triple modular redundancy (TMR), in which three copies of each message bit are created. Triple modular redundancy has $d_{min} = 3$, thus it is able to detect up to two erroneous bits *or* can correct one erroneous bit. The '*or*' in the previous sentence is emphasized because if the TMR circuit is configured to detect a 2-bit error, it cannot correct a 1-bit error; in the 2-bit configuration, the system is unable to tell the difference between a 1-bit error and a 2-bit error, thus majority voting cannot be used. The probability of two errors occurring in a single transmission is much smaller than the probability of a single error, so the 1-bit configuration with majority voting is commonly used.

TMR requires exceptionally large overhead; if a parallel approach is used, each wire is tripled, and the overhead is in terms of area and energy; if a serial approach is used, each wire sends the same information three cycles, and the overhead is in terms of throughput and energy. The information capacity of the code can be defined in terms of a code rate R, which is the ratio of message bits to coded bits. For example, the code rate of TMR is one-third because three bits are transmitted for every one message bit. TMR is generally only used for very small, very important systems; in this work, the one-wire serial communication link connecting the correction systems in the transmitter and receiver is protected with TMR.

7.2.2 Hamming Codes

Hamming codes are more complex to encode and decode than TMR, but have a higher code rate and the same d_{min} as TMR codes ($d_{min} = 3$). In this work, the Hamming code is used for transient error protection (in a 1-bit error correction configuration).

Hamming codes can also detect and correct permanent errors, but each permanent error reduces d_{min} by 1, reducing the code's capability to tolerate transient or intermittent errors. For example, if one wire becomes permanently erroneous, the Hamming code will continuously correct the erroneous bit value every cycle; however, if an additional transient error occurs, the Hamming code will fail because it is unable to correct two errors. Codes which can detect and correct multiple errors (e.g., Bose–Chaudhuri–Hocquenghem (BCH) codes) have large power and area overhead costs, motivating the need for a different type of solution to handle permanent errors.

A simple way to think of the Hamming code is like throwing a ball into a bin. The ball is the message you want to transmit, and the way you transmit the information is by throwing it into a specific bin out of a row of bins representing every possible combination of bits. Without using any coding, it is relatively likely that the ball will wind up getting thrown into the wrong bin (an error). When you use coding, the number of bins increases and now the bins on each side of your target bin funnel the ball into the target bin (the message is turned into a codeword having some redundant bits which get turned back into the original message after transmission). So, the ball winds up in the correct bin even if your aim is a little off (if one bit in your codeword is flipped during transmission).

More formally, the transmitted message m of length k bits is encoded into a codeword c of length n bites by matrix multiplication with a parity check matrix H (of dimension $(n - k) \cdot n$). During transmission, an error vector e is added to the codeword, creating the vector $u = c + e$ which arrives at the receiver. An error syndrome (a representation of potential errors) is calculated by multiplying u by the transpose of the H matrix. The syndrome is an index that can be used to determine the error vector e. The correction is performed by adding e back to u, eliminating the error from the received dataword.

7.2.3   Interleaving

Interleaving is a method of distributing burst errors into multiple codewords to maximize the correction capability of the codes. In the system presented in this chapter, the link is divided into four sections, and each section is coded using a Hamming code. The bits are interleaved (i.e., if the four Hamming codes are labeled $A_0A_1A_2\ldots$, $B_0B_1B_2\ldots$, $C_0C_1C_2\ldots$, $D_0D_1D_2\ldots$, the interleaved output becomes $A_0B_0C_0D_0A_1B_1C_1D_1\ldots$). If a two-bit burst error occurs (a burst error is an error affecting two or more adjacent bits) it will cause one error in two codewords, which can be corrected by the Hamming codes, rather than two errors in one codeword, which cannot be corrected by the Hamming code.

7.3   Related Work

The use of spare modules to replace erroneous ones, especially in array structures, is a long known fault tolerant approach [14]. Spare cells and wires have been used in field programmable gate arrays to bypass defective components [15, 16]. Refan et al. used spare wires to recover from switch failure by connecting each processing element in a network-on-chip (NoC) to two switches [17]; if a permanent fault occurs in one switch, both processing elements share the working switch, and the system reroutes its data accordingly. Grecu et al. have analyzed the use of spare wires in networks-on-chips [18] to increase manufacturing yield; reconfiguration of the links used crossbar switches with redundant channels. Unfortunately, the authors did not discuss the error detection procedure. In another work, Grecu et al. presented a built-in self-test methodology for NoC interconnects [19] and thoroughly discussed manufacture testing methods for NoCs, but they did not address runtime failures. Reick et al. *discuss* dynamic I/O bitline repair using spare wires [20], but their detection and correction processes are not specified.

Run-time reconfiguration has been presented by Lehtonen et al. [21], in which spare wires and a syndrome-storing-based detection (SSD) method were used in an asynchronous system. That prior work has a few significant limitations; for example, the system can tolerate only one permanent error per code interleaving section, and the data flow must be stopped for reconfiguration. In contrast, the in-line test (ILT) method proposed in this work handles as many permanent errors as there are spare wires; further, it allows the link to be reconfigured without interrupting data transmission. The presented reconfiguration system uses a synchronous design methodology and interleaved Hamming codes to achieve higher throughput than the asynchronous design, which had to request retransmissions whenever errors were detected.

7.4 Permanent Error Correction

Permanent error correction in on-chip links using spare wires is a two-step process. First, the permanent error must be detected; then, the link must be reconfigured to avoid transmitting over the faulty wire. The proposed adaptive link framework is presented in Fig. 7.1, and consists of a transmitter, a link, and a receiver. The incoming k-bit wide dataword is encoded in the transmitter to a codeword of width n, which is transmitted through the link and decoded in the receiver.

The decoder is responsible for correcting any errors and outputs the original k-bit dataword. A number of spare wires s have been inserted into the system. *Reconfiguration units* at the transmitter and receiver determine which of the $n + s$ lines carry data and which are left idle. The reconfiguration control units pass reconfiguration information between the receiver and transmitter and synchronize reconfiguration. The *test pattern generator* (TPG) block and test inputs (*test_in*) produce test signals which are received and analyzed in the *error detection and reconfiguration central control unit* to detect permanent errors and initiate reconfiguration.

The reconfiguration process and the remainder of the system framework will be discussed in more detail in Sect. 7.6. We apply our techniques to permanent and intermittent errors in the *link*; the logic units are assumed to function correctly.

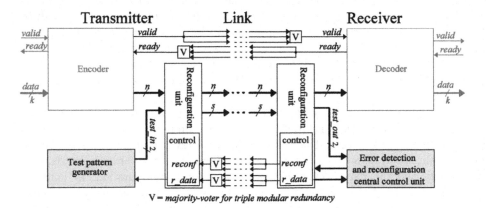

V = majority-voter for triple modular redundancy

Fig. 7.1 The reconfigurable link system

7.5   In-Line Test (ILT) Error Detection Method

The proposed method sequentially routes data from each pair of adjacent wires to a set of available spare wires, allowing each pair to be tested for intermittent and permanent faults. This is achieved during normal operation, without interrupting data transmission. To protect against permanent errors at runtime, the ILT is run periodically, with a period that can be shortened to improve error resilience or increased for energy efficiency.

7.5.1   In-Line Test Procedure

A more detailed diagram of the ILT system is shown in Fig. 7.2. To begin the test, the first test location (i.e., the first wire in the bus) is given to the *Addressing Unit* in the receiver. Three register locations are accessed and provided to the test selection unit: the wire status of the test location (*wireA_status*), the status of the next adjacent wire (*wireB_status*), and the status of the last wire on the link (*spares_left*, which indicates how many spare wires remain available for rerouting).

The *Test Selection Unit* uses these inputs to determine whether a one- or two-wire test should be performed. The *Reconfiguration Control Unit* then determines which, if any, of the two adjacent wires need to be rerouted to spare wires to perform that test, and the link is configured appropriately. Once the reconfiguration is completed, the *Test Pattern Generator* issues a series of test patterns to the wires under test using the *test_in* signal. The received *test_out* signal is compared to a look-up table (LUT), described in Sect. 7.5.2, to determine if there is a permanent error in the wires under test. The LUT indicates which lines need to be flagged as erroneous, and updates the wire status registers appropriately. Functioning wires are reconfigured to carry data once again, and the process is repeated for the each pair of wires (i.e., the test is shifted from wires 1 and 2 to wires 2 and 3, etc.). During each round of tests, wires that were flagged as faulty are retested to prevent intermittent errors from wasting wire resources.

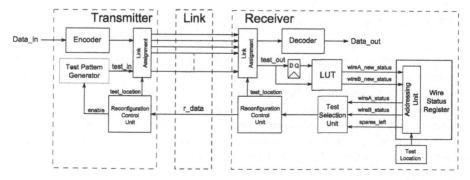

Fig. 7.2 Expanded view of the ILT system

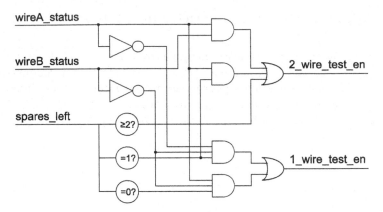

Fig. 7.3 Test selection unit for the ILT system

The ILT system uses a two-wire test to check for shorts between adjacent wires and opens in each wire; however, if only one spare wire remains, the system only reroutes a single wire and performs a single wire test. The single wire test can detect opens in the line but cannot detect a short between the wire under test and its neighbors. If a wire adjacent to the wire under test has been disabled from a previously detected error, the ILT unit will perform the two-wire test on that pair. If no spare wires remain, the system will still periodically re-test each disabled wire in an effort to recover from intermittent errors.

The logic for determining the appropriate test is shown in Fig. 7.3. The *2_wire_test_en* signal is driven high (indicating a two-wire test will be performed) if (1) two or more spares remain, (2) one spare remains and wire A is already flagged as faulty (meaning wire B can be rerouted), or (3) both wires are flagged. A one-wire test will be performed if: (1) one spare remains and neither wire is flagged or (2) no spares remain and wire A is flagged. If no spare wires remain and neither wire is flagged, then no test is performed and the system moves on to the next group.

7.5.2 Analysis of Open and Shorted Wires

Here we analyze the impact of open and shorted interconnect faults on circuit operation to determine the test patterns which will be used in the test procedure. Based on these patterns, a look-up table is created to determine if wires are erroneous using the *test_out* signal. The simulation schematics in Fig. 7.4 were used to analyze the behavior of opens and shorts on a link, with an open simulated using an open resistance, R_{open}, between repeaters, and a short circuit simulated using a short resistance, R_{short}, between two adjacent lines.

Values for the link resistances, R, substrate capacitances, C_{sub}, and metal to metal capacitances, C_{M6}, are taken from an STMicroelectronics 90 nm technology

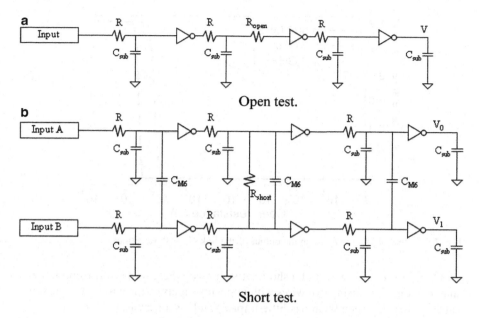

Fig. 7.4 Schematics for wire fault analysis. (**a**) Open test. (**b**) Short test

report. The link resistances and capacitances were computed for 200 m wire sections using the sixth metal layer, M6.

The expected output of a broken wire is fixed regardless of the input value (ignoring coupling effects); however, the failure mode of the wire as its resistance increases (e.g. because of electromigration) is interesting, and highlights an important feature of the ILT system. The output of the open wire, labeled V in Fig. 7.4a, depends on R_{open} and the system operating frequency. To analyze the behavior of a faulty wire, R_{open} was varied between 1 Ω and 100 MΩ. A plot of the resistance vs. output voltage at the receiver is shown in Fig. 7.5 for a variety of operating frequencies.

The plots in Fig. 7.5 all measure the output voltage of a falling transition on the receiver end of the link; the wire is considered faulty at resistances where the output voltage does not pull down to logic low. This is shown to occur at 3 MΩ for the 500 MHz case, with faster operating frequencies failing at lower resistances. This analysis points out that if testing is not at-speed [22], a delay error might not be detected by the test even though the wire will fail at the actual link frequency. (e.g., if the operating frequency is 2 GHz but the test is only run at 500 MHz, an R_{open} of ~1 MΩ will cause an error during operation but will not be caught by the test structure).

The ILT method is an at-speed test, with each test pattern transmission occurring in a single cycle, so these issues are avoided. The ILT implementation tests for open circuits by testing both the 0 and 1 states of each line, detecting both stuck-at-0 and stuck-at-1 faults.

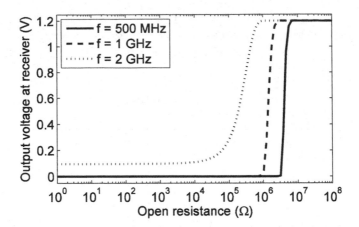

Fig. 7.5 Effect of varying R_{open} on the output of the link ($1 \rightarrow 0$ transition)

To determine the impact of a short between two wires, we perform simulations, shown in Fig. 7.4b, using two wires with opposing inputs (if the wires have the same value, a short between them has little impact) and a short resistance R_{short}.

The results from this simulation are shown in Fig. 7.6, where the impact of a short between two wires is shown to depend on both the β ratio (the ratio of the PMOS and NMOS widths, W_p/W_n) of the repeaters and the value of the resistance. The voltages shown in the figure are taken at the receiver end of the wires, labeled V_0 and V_1 in Fig. 7.4b, to illustrate the impact of the short at the receiver. Figure 7.6a shows the wire whose value would be 0 V without the short; this output is faulty when its voltage is greater than the threshold voltage (~400 mV in the 90 nm technology). Figure 7.6b shows the wire whose value would be 1.2 V without the short; this output is faulty when its voltage is less than the threshold voltage.

If $R_{short} > \sim 10$ kΩ, there are no faults in the link regardless of the value of the β ratio (i.e., the wires are not considered to be shorted). When $R_{short} < \sim 10$ kΩ, the β ratio affects the output response, making detection of short-circuit conditions data dependent. For example, when $\beta = 3$ the output V_1 (shown in Fig. 7.6b) is correct regardless of the short resistance, because the PMOS device in line V_1 dominates that of the NMOS device in line V_0 and the output does not get pulled down. However, in Fig. 7.6a we see that the short causes the other output (V_0) to pull up to 1.2 V incorrectly for resistances below ~10 kΩ.

Thus, for a short circuit to be detected at the output of a wire V, two conditions must be met. The two shorted wires must have different values, *and* the input of V must be the value dominated by the short (i.e., from Fig. 7.6, an error occurs when $\beta > \sim 2.5$ only if $V_0 = 0$, and an error only occurs when $\beta \leq \sim 2.5$ if $V_0 = 1$).

The potential for a short to only affect the output under certain conditions requires two test cases to evaluate whether a short exists between two wires. Each combination of contention currents must be examined (i.e., the case where $A = 1$ and $B = 0$, as well as the case where $A = 0$ and $B = 1$). These two tests are also

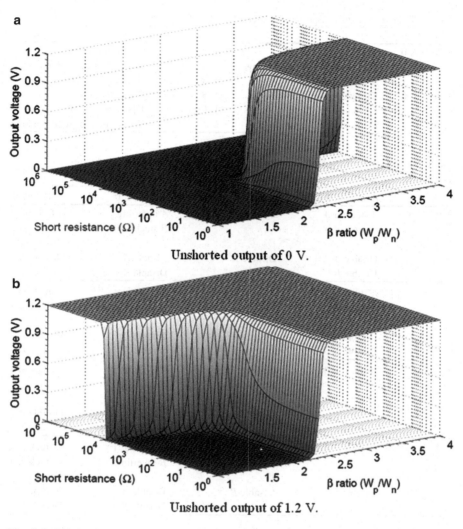

Fig. 7.6 Effect of varying short resistances and beta ratios on the output of two wires with opposing states. (**a**) Unshorted output of 0 V. (**b**) Unshorted output of 1.2 V

capable of detecting an open circuit, as both values (0 and 1) of each line are examined. Each test has four possible output combinations, listed in Table 7.1 along with potential causes of each output response.

The output from the two test cases can result in 16 possible combinations, listed in Table 7.2 along with potential causes of each response and the necessary action to take (shown in bold). Rows indicate results from Test 1 ($A = 1$, line $B = 0$), while columns indicate results from Test 2 ($A = 0$, line $B = 1$). For example, the box where the result of Test 1 is (1,0) and the result of Test 2 is (0,1) represents the condition where the outputs of A and B behave correctly, thus no wires need to

Table 7.1 Possible output responses for each of the two fault tests

Test 1: $A = 1, B = 0$		Test 2: $A = 0, B = 1$	
$A, B = 0, 0$	Open on Line A	$A, B = 0, 0$	Open on Line B
	Wires shorted, B dominating		Wires shorted, A dominating
$A, B = 0, 1$	A and B open	$A, B = 0, 1$	Correct
$A, B = 1, 0$	Correct	$A, B = 1, 0$	A and B open
$A, B = 1, 1$	Open on Line B	$A, B = 1, 1$	Open on Line A
	Wires shorted, A dominating		Wires shorted, B dominating

Table 7.2 Potential effects and necessary actions to correct open and shorted wires

Test 1	Test 2			
A, B	$A, B = 0, 0$	$A, B = 0, 1$	$A, B = 1, 0$	$A, B = 1, 1$
0, 0	A stuck at 0	A stuck at 0	A inverting	A follows B
	B stuck at 0		B stuck at 0	
	↓	↓	↓	↓
	Disable A	Disable A	Disable A	Disable A
	Disable B		Disable B	
0, 1	A stuck at 0	A stuck at 0	A inverting	A inverting
	B inverting	B stuck at 1	B inverting	B stuck at 1
	↓	↓	↓	↓
	Disable A	Disable A	Disable A	Disable A
	Disable B	Disable B	Disable B	Disable B
1, 0	B stuck at 0	Correct Functionality	A stuck at 1	A stuck at 1
			B stuck at 0	
	↓	↓	↓	↓
	Disable B	Do nothing	Disable A	Disable A
			Disable B	
1, 1	B follows A	B stuck at 1	A stuck at 1	A stuck at 1
			B inverting	B stuck at 1
	↓	↓	↓	↓
	Disable B	Disable B	Disable A	Disable A
			Disable B	Disable B

be disabled. There are two possible fault modes, including stuck-at faults, possibly from a break in the wire, or inverted response, which may be a result of delay errors or a combination of a broken wire and a short to another wire.

7.6 Adaptive Correction Framework

In the following subsections, the structure and operation of the modules from Fig. 7.1 (aside from the previously described error detection units) are presented.

7.6.1   Encoder and Decoder

The encoder and decoder provide protection against transient faults. The encoder calculates check bits which are transmitted together with the data, and these check bits are used in the decoder to detect and correct errors. The reconfiguration system is designed as a separate layer from the underlying data transmission. This means that there are no error control coding requirements for the reconfiguration system to be able to bypass permanent errors.

The system uses *valid* and *ready* signals to indicate when the data is valid for processing and when the receiver is ready for new data. These signals are needed to handle situations when there is no data to be transmitted, or when the receiver cannot store new data words because it is out of buffering space (e.g., because of network congestion). As mentioned earlier, these control signals are protected using triple modular redundancy (TMR).

7.6.2   Reconfiguration Units

The function of the reconfiguration unit is to route the data around the erroneous wires or wires under test. To balance the delay within routed wires, the reconfiguration ripples through the bus as shown in Fig. 7.7.

The number of spare wires s to be inserted into the system depends on the probability of a permanent error in a wire, the number of wires and the desired link lifetime. The number of spare wires has an effect not only on the permanent error tolerance but also on the complexity of the reconfiguration units. Each additional spare wire inserts some logic: the selection logic for the added spare, and one new input to select from for each wire in the link (the total number of inputs per wire is $s + 2$, including the data input, s data inputs for rerouting, and the ILT test input). In addition, the total number of control registers per wire is $\lceil \log_2(s+1) \rceil$. The impact of s on area and energy is further analyzed using a case study, and will be discussed in Sect. 7.7.

The spare wires have two purposes—to replace faulty wires, and to temporarily bypass in-use wires to allow them to be tested. For the latter purpose, spare wires can be used to replace in-use wires for testing. The retest property allows bypassed wires to be restored if their errors turn out to be intermittent rather than permanent.

The *test_in* and *test_out* signals are only implemented with the ILT detection method. The method tests two adjacent wires at a time, so the reconfiguration unit at the transmitter must be able to connect the two test inputs to the correct pair of spare wires from s available spares. Similarly, at the reconfiguration unit at the receiver the two corresponding test outputs are selected from the set of outputs. The input selection is implemented using a control signal of width $\lceil \log_2(s-1) \rceil$

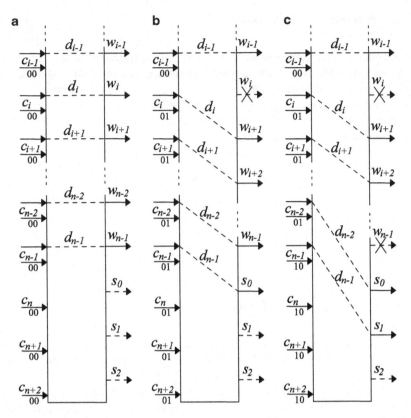

Fig. 7.7 The reconfiguration unit: (**a**) no permanent errors, (**b**) a permanent error at location i, (**c**) permanent errors at locations i and $n-1$

(there are $s-1$ pairs of adjacent spare wires). At the receiver there is a test control input which provides the address of the first test output, i. The other test output is set to address $i+1$.

7.6.3 Reconfiguration Unit Control

The error detection circuits in the *Error detection and reconfiguration central control unit* provide the location of any erroneous wires. To bypass these wires, the control sections of the reconfiguration units at each end of the link are used to transmit the reconfiguration information from the receiver to the transmitter, as well as synchronize the reconfiguration so that both the transmitter and receiver do the reconfiguration in the same cycle. The transmission protocol is presented in the following subsection.

Since the spare wires are shared by all the wires in the link (in contrast to [21], where each spare was fixed to a section of the link), the control portions of the reconfiguration units must be aware of the current state of each wire in the link. The request for a new reconfiguration is combined with this state information to produce matching control signals at the transmitter and receiver. To ensure that the system has a minimal effect on the data flow, there is a strict requirement for how much processing can be done during the reconfiguration cycle. Therefore, the control unit provides a separate control signal for each wire (width$\lceil \log_2(s+1) \rceil$), reducing the amount of logic inserted into the critical path. The control signal provides the number of reconfigurations which have occurred at all indexes less than or equal to the wire address, so one signal can be directly used for controlling the wire routing. If wire i is erroneous, the control values of wire i and wire $i-1$ will be different (in the case of $i=0$ the wire is indicated as erroneous if the control signal is non-zero). Figure 7.7 illustrates the usage of control signals, shown just below each control input. In this example there are three spare wires. Figure 7.7a shows the case where there are no erroneous wires, thus the data at each wire location is passed straight through the reconfiguration unit, and all control signals are set to '00'. Figure 7.7b shows the case where wire w_i is erroneous. Once detected, this causes the control signal at all locations greater than or equal to i to be set to '01', which in turn causes data passing through those wires to be shifted by one location. Figure 7.7c shows the effect of an additional wire error, with both wires w_i and w_{n-1} marked erroneous. The control signals at all locations greater than or equal to $n-1$ are set to '10', which causes the data at all locations greater than or equal to $n-2$ to be redirected (d_{n-2} was already rerouted to w_{n-1}, so it is shifted a second time).

The reconfiguration control unit has two modes of operation. It can mark a wire erroneous, or remove the mark from a wire. The latter is needed to return a wire to normal operation if the error on it is intermittent instead of permanent. The mode can be provided as an input to the reconfiguration unit or it can be extracted from the reconfiguration control signals. The latter method is used in the transmitter to minimize the control transmission length from receiver to transmitter, while the former is used in the receiver because the information is already available there. Control values are computed by incrementing or decrementing the control value of each wire based on the reconfiguration location and mode.

7.6.4   Transmission Protocol

To minimize link area and energy, the transmission of control data is serial, using only one signal r_data, which is protected with TMR (see Fig. 7.1). Synchronization is achieved using the *reconf* signal (also protected by TMR), which enables the transmission of control data at the appropriate time. The transmission protocol is divided into three phases, including: (1) The initialization of a transmission, (2) the transmission of the error location err_loc $\lceil \log_2(n+s+1) \rceil$ bits; the all-ones location is

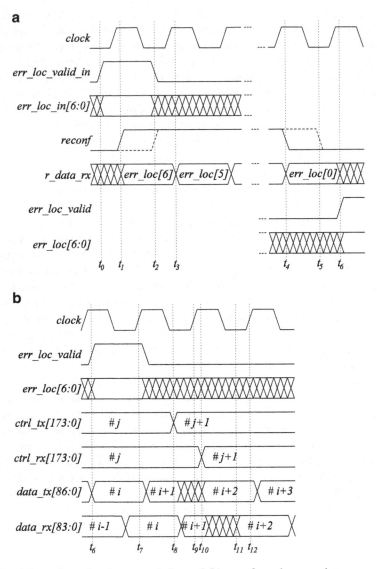

Fig. 7.8 (**a**) Reconfiguration data transmission and (**b**) reconfiguration procedure

reserved for the purpose of starting a test without doing any reconfiguration), and (3) the end of the transmission. The reconfiguration procedure is activated immediately after the transmission.

Figure 7.8 illustrates the reconfiguration procedure and the transmission of reconfiguration information from receiver to transmitter. In this example *err_loc* is seven bits wide. The transmission protocol and reconfiguration procedure are divided into timesteps labeled below the waveforms to aid the following description.

Phase 1: The reconfiguration procedure is initiated by the *reconfiguration central control unit* at the receiver, which loads the reconfiguration location to *err_loc_in* and sets *err_loc_valid_in* (t_0 in Fig. 7.8a). The *reconfiguration unit control* at the receiver detects the *err_loc_valid_in* signal at the next clock edge, stores the error location, forwards its uppermost bit (*err_loc*[6]) to *r_data* and sets *reconf* high (t_1). Setting *reconf* signals the start of a new transmission. In Fig. 7.8a, the solid line for *reconf* corresponds to transitions at the receiver and the dashed line corresponds to transitions at the transmitter one link delay later (t_2). For *r_data* only the receiver transitions are shown, indicated by *rx* at the end of the signal name.

Phase 2: The reconfiguration control unit at the transmitter latches the information on the rising clock edge and the transmission proceeds to the next error location bit (t_3 in Fig. 7.8a). The transmission proceeds one bit at a time, so the length of the total transmission depends on the width of the error location vector.

Phase 3: After transmission of the second to last error location bit *err_loc[1]* (t_4 in Fig. 7.8a), the reconfiguration control unit at the receiver resets *reconf*. When this transition is seen at the transmitter (t5), the end of transmission has been reached. The error location *err_loc* is forwarded to the control signal calculation, and the calculation is triggered by *err_loc_valid* at the next rising clock edge (t_6 in Fig. 7.8a). The receiver side does the same with its reconfiguration control unit (t_6).

The reconfiguration procedure is illustrated in Fig. 7.8b. Note that the data transmission is uninterrupted by the reconfiguration process. This is achieved by taking advantage of the delay of the link using wave pipelining. The *err_loc_valid* signal is detected at the transmitter on the rising clock edge (t_7 in Fig. 7.8b). On the falling edge of that clock pulse, the transmitter side is reconfigured (t_8). The reconfiguration is carried out by updating the control register at the transmitter reconfiguration unit *ctrl_tx*. The receiver latches the data (t_9 in Fig. 7.8b) before the new values from the reconfiguration propagate through the wire. This provides a timing constraint defined as

$$t_{R\_TX} + t_L + t_{R\_RX} \geq \frac{T_{clk}}{2} + t_h \qquad (7.1)$$

where t_{R_TX} is the shortest delay through the reconfiguration logic at the transmitter, t_L is the delay through the link, t_{R_RX} is the shortest delay through the reconfiguration logic at the receiver, T_{clk} is the clock period and t_h is the hold delay of the input register at the receiver.

The reconfiguration at the receiver is then completed by updating the control register *ctrl_rx* at the next rising clock edge (t_{10} in Fig. 7.8b). The reconfiguration must be completed before the next data arrives at the receiver (t_{11}) and is latched at the next rising clock edge (t_{12}).

7.7   Case Study and Comparison with Prior Work

Hamming codes are the most widely used codes in previous research on interconnect
link error protection [2–6, 8]. As mentioned earlier, the minimum distance of a
Hamming code is 3, so it can correct a single error in each codeword. Interleaving
a large codeword into multiple groups of smaller Hamming codes can enhance the
burst error tolerance of the larger codeword and also tolerate multiple errors
(provided each error occurs in a separate group) [8, 23]. Because of its popularity,
Hamming coding with interleaving is chosen as a reference system for comparison,
and is also used in the presented ILT system. The presented system is also compared
to a syndrome-storing-based detection (SSD) method, which is an alternative permanent error correction method in which the syndrome (the detected error pattern) of
consecutive codewords are compared to check if an error lasts multiple cycles. Unlike
the ILT method, the SSD method cannot reclaim a wire once it is disabled—
permanent errors can be bypassed, but intermittent errors cannot be detected.

For the case study, we set the data width $k = 64$ bits, and select a shortened
Hamming (21,16) code with four interleaving sections (a shortened Hamming code is
a code in which some message bits are removed to shrink the message size to a
desired bit width—in this case a message width of 16 is used), resulting in a codeword
width $n = 84$. The system can correct all single errors, up to 77% of all double errors
and burst errors affecting up to four adjacent wires [23].

The systems are also compared to a more complex coding approach, capable of
detecting multiple errors. The *Bose–Chaudhuri–Hocquenghem* (BCH) codes are a
class of linear block codes that can be easily constructed according to specifications
for correcting as many errors as is required. We construct a code capable of
detecting three errors and shorten it to match the dataword width, resulting in a
shortened BCH (85,64) code.

For this case study we use three spare wires ($s = 3$), which can be controlled with
a two-bit signal as was explained in Sect. 7.4. The four systems created for
comparisons and analyses are listed in Table 7.3, where t_{tran} and t_{perm} are the transient
and permanent error correction capabilities respectively.

7.7.1   Implementation Results

The analysis was performed by synthesizing each design in STMicroelectronics
90 nm technology. The wire area is not included in the result values, but their effect
on timing and power consumption is taken into account.

Table 7.3 Permanent error correction case study systems

Name	Code	$n + s$	t_{tran}	t_{perm}
Ham	4 × Hamming (21,16)	84	1*	
ILT	4 × Hamming (21,16)	87	1*	3
SSD	4 × Hamming (21,16)	87	1*	3
BCH	BCH (85,64)	85	3	

Table 7.4 Characterization of the case study systems

Circuit area (μm^2)

Name	Encoder	Decoder	Total	Gate equiv.
Ham	5,454	7,772	13,226	3,013
ILT	22,420	27,745	50,165	11,427
SSD	18,859	27,692	46,551	10,604
BCH	6,762	68,413	75,176	17,124

Clock frequency (GHz), throughput (MWord/s), and latency

Name	Clock freq.	Throughput	Latency Clock cycles	ns
Ham	1.8	1,800	4	2.22
ILT	1.4	1,400	4	2.86
SSD	1.4	1,400	4	2.86
BCH	1.1	1,100/367[a]	5/8[a]	4.55/7.27[a]

Energy per transmitted flit (pJ)

Name	Encoder	Decoder	Link	Total
Ham	7.94	9.98	54.60	72.52
ILT	10.67	13.73	54.60	79.00
SSD	10.23	14.61	54.60	79.44
BCH	9.90/15.58[a]	29.50/151.94[a]	55.25	94.65/222.76[a]

Stand-by power (mW)

Name	Encoder	Decoder	Total
Ham	0.84	0.75	1.59
ILT	0.81	1.19	2.00
SSD	0.88	1.12	2.00
BCH	0.49	1.75	2.24

[a]no errors/with errors

The delay and energy values for the link are from a study [24] where a link was modeled as an RC line of length 2 mm, using metal layers 5 and 6 in STMicroelectronics 90 nm technology at 1.2 V. The wires have minimum width, spacing to adjacent wires is double the minimum width, and the repeater spacing was optimized to minimize latency [25]. With average crosstalk, the delay and energy per wire per transmission, including the link drivers and repeaters, are 260 ps and 1.3 pJ, respectively. Average values for crosstalk are provided using a recently presented technique that averages crosstalk effects without affecting other link parameters [26].

The timing margin was set to 25% to account for sources of variation, so the circuits were synthesized at a clock frequency 25% higher than that reported in Table 7.4. For instance, ILT and SSD can operate at 1.75 GHz in normal operating conditions, but the results are reported for frequencies of 1.4 GHz.

The sum of the link delay, flip-flop CLK-Q propagation delay, and flip-flop setup delay provides the minimum propagation delay through the link, which, with the timing margin included, results in a maximum frequency of 1.8 GHz. If a higher frequency is required, the link can be pipelined, or different design parameters, such as wider wires or larger wire spacing, can be used.

The reconfiguration logic delay is 130 ps including both the units at the transmitter and receiver. Low delay overhead for the critical path is achieved by constructing the multiplexers from tri-state buffers. The additional delay of 130 ps requires the frequency to be lowered by 22% to 1.4 GHz (including the 25% design margin). The majority-voter delay for triple modular redundancy in the control lines is less than the delay of reconfiguration logic.

Area, speed, energy, and power results are presented in Table 7.4. Each system is pipelined to achieve high throughput. The transmitters are one pipeline stage, the link is another stage, and the receiver requires two stages for each system other than BCH. BCH decoding has two modes, one when there are no errors in the link, and the other when errors need to be corrected. Implementation details for the BCH realization are presented in [23].

Area values include each system component and the registers required for pipelining. Note that the transmitter area overhead for the BCH is smaller than the transmitter area overhead of the ILT and SSD systems. This is because the SSD and ILT systems require large configuration units in the transmitter to handle permanent errors, while the majority of the overhead in the BCH implementation is in the complex decoder unit. Energy values were calculated from the average power and simulation time of passing 10^4 random datawords through the systems at maximum throughput. The Ham, ILT and SSD system results include a transient error during transmission. BCH results are reported both with and without errors.

The implementations use clock gating to reduce power and energy consumption. This can be seen from the stand-by power consumptions, which is measured when no data is flowing through the system (input signal *valid* is low) but the clock is not stopped. Stand-by power consumptions are only a few milliWatts, a small fraction of the values during operation. Note that the power values are reported at the operating frequency of each system.

The energy consumption, area and latency are compared in Fig. 7.9, where each is presented relative to the value of the reference circuit Ham. The values for BCH are for the mode with errors to reflect the situation with permanent errors for which the systems are designed. From the figure it can be seen that the area of the ILT and SSD systems is 3.5–3.8 times as large as that of the Hamming reference circuit, but still only two thirds of the area required by the BCH implementation. The energy and latency overheads of the reconfigurable systems are 10% and 29% respectively as compared to the Hamming approach, and less than 40% of the BCH realization (36% energy and 39% latency).

One reconfiguration procedure as described in Fig. 7.8 consumes 99 pJ, which is about 1.3 times the energy consumed by one data transfer. In the SSD system, the energy overhead is negligible since it is only needed when an error is detected, and the maximum number of reconfigurations is the number of spare wires. In the ILT system, reconfigurations are done more frequently. One complete test round requires $2(n + s)$ reconfigurations, and $n + s - 1$ test rounds each consisting of two test patterns.

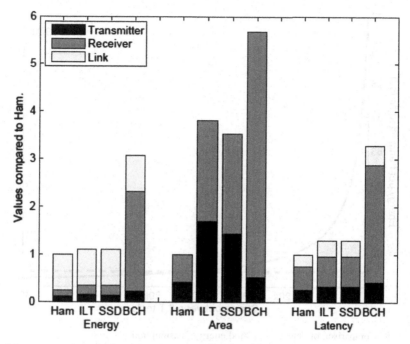

Fig. 7.9 Case study comparison of energy, area and latency

The total amount of energy required to run one ILT test round is 20.3 nJ, of which 7.1 nJ is consumed at the transmitter, 8.8 nJ at the receiver and 4.3 nJ in the link transmitting the reconfiguration data and the actual testing. One test round takes 2,257 clock cycles so the energy overhead per transmitted flit is 9.0 pJ. When each round of testing is separated by a number of cycles, the energy overhead per transmitted data flit is reduced. This is shown in Fig. 7.10, where the energy per transmitted data flit is shown as a function of the number of data cycles between ILT test rounds. For comparison, the SSD energy value is shown. At the intersection of the two lines, 44,129 cycles, both methods have the same energy consumption per transmitted flit. This means that if the in-line test is run less frequently than every 46,386 transmitted words, it consumes less energy than the SSD system.

7.7.2 Error Tolerance Analysis

The error correction capability of a coding approach can be described by the probability of a correct transmission in the presence of err errors, P_{err}. This can be refined to take into account the combined probabilities of correct transmission under transient ($tran$) or permanent ($perm$) errors, P_{tran} and P_{perm} respectively.

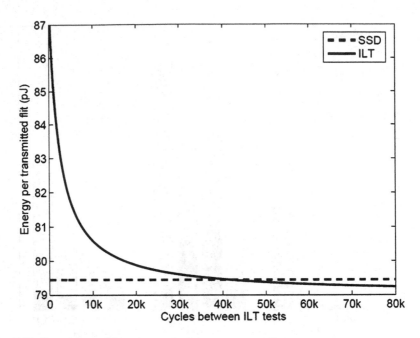

Fig. 7.10 Comparison of detection method energy consumption

The combined probability is defined as P_{tran}^{perm} which represents the probability of correct transmission in the presence of *tran* transient errors and *perm* permanent errors.

Since Ham and BCH systems do not distinguish between the two types of errors, the combined probability of a correct transmission under both transient and permanent errors is

$$P_{tran}^{perm} = P_{tran+perm} \tag{7.2}$$

where $P_{tran+perm}$ is the probability of a correct transmission in the presence of the total number of transient and/or permanent errors.

Because each type of error is addressed separately in the ILT and SSD systems, the combined probability of correct transmission in those systems is

$$P_{tran}^{perm} = P_{tran} \, P_{perm} \tag{7.3}$$

as long as the time period between when a permanent error occurs and when it is corrected is much less than the mean time between transient errors (If a permanent error occurs at time t and is not corrected until time $t + a$, the time a should be much less than the time between transient errors). With the presented detection delays for the ILT system, the above is true for transient bit error rates less than 10^{-6} errors/cycle. For the ILT system $P_{perm} = 1$ when $perm \leq s$.

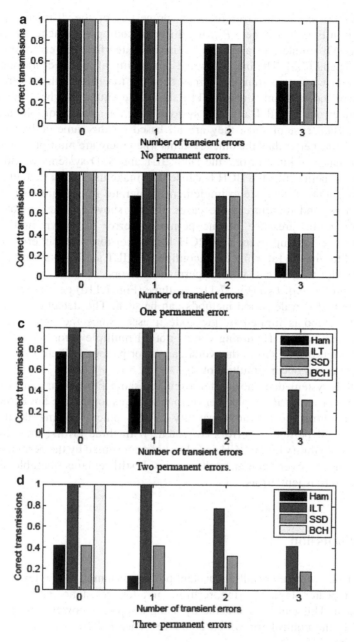

Fig. 7.11 Comparison of error correction capability of the case study. (**a**) No permanent errors.
(**b**) One permanent error. (**c**) Two permanent errors. (**d**) Three permanent errors

A comparison of the error recovery properties of the different systems is illustrated in Fig. 7.11, where P_{err} for different coding approaches is calculated via MATLAB simulations and P_{tran}^{perm} is calculated for each error combination using (7.2) and (7.3). The bars show the probability of a correct transmission in each system when 0–3 transient errors occur in the link, with all errors being assumed as independent. Figure 7.11a shows the situation where there are no permanent errors. The ILT and SSD systems have the same error tolerance as the reference Ham system since they are all based on the same coding approach. BCH performs better than the other systems when there are multiple single errors. In the presence of burst errors, the Ham, ILT and SSD systems would perform similar to or better than the BCH because they make use of interleaving. For our case study, none of the systems implemented tolerates four or more simultaneous single errors, and therefore those cases are not shown in the graphs. System reliability in the presence of one permanent error is shown in Fig. 7.11b. The reference systems Ham and BCH lose a portion of their error tolerance compared to Fig. 7.11a, while the reconfigurable ILT and SSD systems maintain the same error tolerance as with no permanent errors.

In the presence of two (Fig. 7.11c) or three (Fig. 7.11d) permanent errors, the difference between detection methods can be seen. The detection capability of the SSD method is limited by the code it uses. Since the underlying coding approach is based on Hamming codes, not all multiple error scenarios can be detected. Figure 7.11 shows the worst case error patterns, where multiple single permanent errors occur simultaneously.

The ILT system does not suffer from the same limitations, and can detect all permanent faults and replace erroneous wires as long there are enough spare wires. The error correction capability of the other systems declines with the number of permanent errors corrected. With three errors, the whole error correction capability of the BCH code has been consumed by the permanent errors and it cannot recover from any other errors. It still remains operable as long as there are no transient errors.

7.8   Discussion

An erroneous wire can result in constant power consumption if it leaves the input of some repeater stages floating, resulting in a low-resistance path between V_{DD} and ground. This can be overcome by using a separate power gating circuit for each wire, the control for which could be generated from the reconfiguration control signals.

Spare wires can be efficiently used for shielding if their input is connected to V_{DD} or ground. This can be especially beneficial if some wires are slower than others (making them more susceptible to timing errors). Slower wires could be detected with an enhancement to the presented in-line testing system. Currently the system grounds all unused spare wires.

The decrease in throughput resulting from the reconfiguration 130 ps latency overhead could be avoided by inserting new pipeline stages for the reconfiguration units at the transmitter and receiver. This would mean trading latency, area and energy for higher throughput. The area and energy overhead of this approach could be partially compensated by retiming transmitter and receiver pipeline stages to loosen timing constraints in the synthesis tool.

The total latency introduced by the proposed error protection circuitry is three clock cycles when compared to a link without any protection. The additional cycles are needed for the Hamming encoding and decoding. SSD always requires an ECC to be used on the link; however, ILT can be used on links without any ECC and thus, without additional clock cycles. This could be useful in Networks-on-Chip where end-to-end transient error protection is used (end-to-end transmission involves one-time encoding and decoding even though the data packet may travel multiple hops before reaching its destination), or in media applications where small periods of data corruption are acceptable.

In many cases there are buffers at the inputs and outputs of the link (e.g., buffering at the routers of a NoC). This buffering capacity can be replaced by the buffer stages introduced to the transmitter and receiver when an ECC is used. Using these pre-existing buffers, we would not need any additional clock cycles and the latency increase would be eliminated.

The area overhead of the circuits should be considered in context. For instance, in a mesh-shaped NoC with bidirectional links, there are five links per node, thus six transmitter/receiver pairs. If a node is of size 2×2 mm [27], the area overhead of the reconfigurable systems is approximately 5%, while the overhead of the BCH would be approximately 9%.

Optimal selection of the number of spare wires s is nontrivial. Increasing s will also increase the delay of the reconfiguration units. The multiplexers have been constructed from tri-state buffers so that their impact when adding spare wires is limited to the delay caused by the output capacitance of the additional tri-state buffer together with the capacitance of the additional wiring. In [18], which concentrated on yield improvement, the number of spare wires could be as high as the number of original wires. For run-time fault tolerance, a smaller number of spare wires is sufficient. The main benefit of the presented approach is to avoid use of complex coding schemes while still having good permanent error tolerance.

The impact of the number of spare wires on the system area overhead and energy consumption was analyzed by synthesizing the ILT design with different numbers of spare wires. The normalized results are shown in Fig. 7.12. The area grows considerably when adding the second or fourth spare wire. This logarithmic dependence was already predicted in Sect. 7.6. The changes in energy consumption are minor. From Fig. 7.12 it can be concluded that the selection of spare wires is mainly a trade-off between reliability and area. Spare wire numbers of 1, 3 and 7 are optimal choices due to the logarithmic dependence of the area.

In the ILT system, the reconfigurations of wires in preparation for the test signals results in the majority of delay and energy consumption. One way to decrease the number of control transmissions and reconfigurations would be to

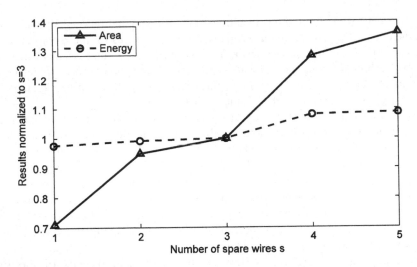

Fig. 7.12 The impact of the number of spare wires s on area overhead and energy consumption

introduce a special control packet to be used between tests to indicate two reconfigurations at once. This could potentially halve the number of control transmissions and reconfigurations.

The combination of the ILT and SSD methods could achieve additional benefits beyond those discussed in this work. If the SSD method were used to detect permanent errors, the ILT properties could be used to further test wires declared erroneous. This would make it possible to return wires to normal usage if an error was intermittent. The SSD could also be used to trigger an ILT test round, thus minimizing the overhead of unnecessary test rounds.

References

1. De Micheli G, Benini L (2006) Networks on Chips: Technology and Tools (Systems on Silicon). Morgan Kaufmann, CA
2. Bertozzi D, Benini L, De Micheli G (2005) Error control schemes for on-chip communication links: the energy-reliability tradeoff. IEEE trans Computer-Aided Design of Integrated Circuits and Syst 24:818–831
3. Li L, Vijaykrishnan N, Kandemir M, Irwin MJ (2003) Adaptive error protection for energy efficiency. Int Conf Computer-Aided Design 2–7
4. Murali S et al (2005) Analysis of error recovery schemes for networks on chips. IEEE Design and Test of Computers 22:434–442
5. Rossi D, Angelini P, Metra C (2007) Configurable error control scheme for NoC signal integrity. 13th IEEE Int On-Line Testing Symp 43–48
6. Sridhara SR and Shanbhag NR (2005) Coding for system-on-chip networks: a unified framework. IEEE Trans Very Large Scale Integr (VLSI) Syst 13:655–667
7. Yu Q, Ampadu P (2008) Adaptive error control for reliable systems-on-chip. IEEE Int Symp on Circuits and Syst 832–835

8. Zimmer H, Jantsch A (2003) A fault model notation and error-control scheme for switch-to-switch buses in a network-on-chip. 1st IEEE/ACM/IFIP Int Conf on Hardware/Software Codesign and System Synthesis 188–193
9. Lehtonen T, Wolpert D, Liljeberg P, Plosila J, Ampadu P (2010) Self-adaptive system for addressing permanent errors in on-chip interconnects. IEEE Trans Very Large Scale Integr (VLSI) Syst 18:527–540
10. Srinivasan GR (1996) Modeling the cosmic-ray-induced soft-error rate in integrated circuits: an overview. IBM J Res and Dev 40:77–89
11. Heidel DF et al (2008) Alpha-particle-induced upsets in advanced CMOS circuits and technology. IBM J Res and Dev 52:225–232
12. Leray JL (2007) Effects of atmospheric neutrons on devices at sea level and in avionics embedded systems. Microelectronics Reliability 47:1827–1835
13. Lin S, Costello DJ (2004) Error Control Coding, 2nd Ed. Prentice Hall, NJ
14. Johnson B (1989) Design and Analysis of Fault Tolerant Digital Systems. Addison-Wesley, MA
15. Hanchek F, Dutt S (1998) Methodologies for tolerating cell and interconnect faults in FPGAs. IEEE Trans on Computers 47:15–33
16. Yu AJ, Lemieux GGF (2005) Defect-tolerant FPGA switch block and connection block with fine-grain redundancy for yield enhancement. Int Conf on Field Programmable Logic and Applications 255–262
17. Refan F, Alemzadeh H, Safari S, Prinetto P, Navabi Z (2008) Reliability in application specific mesh-based NoC architectures. IEEE On-Line Testing Symp 207–212
18. Grecu C, Ivanov A, Saleh R, Pande PP (2006) NoC interconnect yield improvement using crosspoint redundancy. 21st IEEE Int Symp on Defect and Fault Tolerance in VLSI Syst 457–465
19. Grecu C, Ivanov A, Saleh R, Pande PP (2007) Testing network-on-chip communication fabrics. IEEE Trans on Computer-Aided Design of Integrated Circuits and Syst 26:2201–2214
20. Reick K et al (2008) Fault-tolerant design of the IBM Power6 microprocessor. IEEE Micro 28:30–38
21. Lehtonen T, Liljeberg P, Plosila J (2007) Online reconfigurable self-timed links for fault tolerant NoC. VLSI Design 94676:1–13
22. Shin J, Kim H, Kang S (1999) At-speed boundary-scan interconnect testing in a board with multiple system clocks. Design, Automation and Test in Europe 473–477
23. Lehtonen T, Liljeberg P, Plosila J (2007) Analysis of forward error correction methods for nanoscale networks-on-chip. 2nd Int Conf on Nano-Networks 1–5
24. Rantala V (2008) Hybrid mesh-ring network-on-chip with adaptive routing. Master's Thesis, University of Turku, Finland
25. Ismail YI, Friedman EG (2000) Effects of inductance on the propagation delay and repeater insertion in VLSI circuits. IEEE Trans Very Large Scale Integr (VLSI) Syst 8:195–206
26. Kaul H, Seo JS, Anders M, Sylvester D, Krishnamurthy R (2008) A robust alternate repeater technique for high performance busses in the multi-core era. IEEE Int Symp on Circuits and Syst 372–375
27. Dally WJ, Towles B (2001) Route packets, not wires: on-chip interconnection networks. Design Automation Conf 684–689

Chapter 8
Future Work and Open Problems

Although this book has concentrated on the impacts of and solutions to temperature variation, future chip design will require a holistic approach to PVT variation. Process, voltage, and temperature variation each impose fundamental limits on circuit design which are fast approaching and will require a great deal of effort to overcome. The problems arising from the immense integration of nanoscale design will link each type of variation with the others; process variation will result in high-leakage components which will cause extreme local temperature variations; process and temperature variations will affect the amount of current flowing through each device, impacting voltage variations; temperature variation will result in advanced aging and changing process variation profiles, etc.

Process variations may be improved through better fabrication methods, or the effects of process variation can be improved through post-process tuning; however, at the atomic scale, the fundamental quantum behavior of semiconductor devices will result in variations which may become orders of magnitude more important than they are right now; in a given channel there may be zero or two or three dopant ions, and this will require path-level adaptation which is infeasible with current adaptive system design.

Voltage variations may be improved through more robust voltage supplies and DC-DC converters, and the introduction and improvement of decoupling capacitors, but on the nanoscale level switching noise will be a fundamental problem. The voltage on a segment of a supply rail is fundamentally dependent on the data input to the circuit it supplies, and the amount of current drawn by the circuits is again intimately tied with local (and temporal!) variations in process and temperature.

Temperature limitations are currently the biggest road block impeding performance improvement in large-scale systems. As mentioned in Chap. 1, the energy densities of integrated circuits are quickly approaching the levels of a nuclear reactor; system performance is limited by the amount of cooling available, and increased core frequencies will result in larger on-chip temperature gradients—particularly with the growing popularity of SoI and 3D integration.

In this book, we have shown that temperature variation can be extremely important in system design, and presented some systems which will be able to reduce

D. Wolpert and P. Ampadu, *Managing Temperature Effects in Nanoscale Adaptive Systems*, DOI 10.1007/978-1-4614-0748-5_8,
© Springer Science+Business Media, LLC 2012

its impact. These systems are capable of combating the current problems facing temperature variation, but one major limitation of these systems is that future materials, particularly FD-SOI systems, can have temperature variations occurring orders of magnitude more quickly than the detection systems presented in this work. Low overhead sensors will be critical for mapping local temperature variations, but if these variations occur too quickly to be detected, even adaptive systems will require extremely large guardbands to ensure functionality. While process variation guard-bands can be reduced by using frequency binning (grouping parts by their maximum speeds or power budgets), the speed limitations caused by runtime variation guardbands cannot be offset by binning. One alternative is to use the proposed temperature compensation device approach to limit the impact of temperature on device current, but this approach may result in large delay overheads if applied to an entire chip.

To combat these problems, we will need adaptive systems that are capable of quickly responding to changes in runtime conditions. we believe this will be among the largest problems in design of future systems, and the all-digital sensors presented in this paper are among the fastest reported to date for the resolutions mentioned (most temperature sensor designs do not treat speed as a primary, or even secondary concern). This is a two-part problem: faster sensors are critical, but the adaptive systems must be able to react more quickly as well, and the combined delay of sensing and system adaptation must be scaled down significantly to make variation-aware systems useful for future generations of nanoscale system design.

In addition to fast response time, one of the most important ways in which adaptive systems can be improved is through local optimization. Adaptive methods in large systems must support the lowest common denominator, e.g., the slowest component or the most power hungry core. Dividing the system into smaller units will provide a more comprehensive solution, although this must be carefully balanced by overhead costs. As systems become further susceptible to temperature variation (and other types of variation as well), local adaptations will provide greater gains, and it will become beneficial to provide these adaptations at the core level, or even the path level. To accomplish these optimizations, further work is needed on reducing the overhead of the adaptive systems.

The increased susceptibility to PVT variations will also likely increase the use of asynchronous designs, such as locally-asynchronous globally-synchronous (LAGS) or globally-asynchronous locally-synchronous (GALS) approaches. These approaches allow local units to be optimized to complete a task while the global system control is either synchronized or uses handshaking to avoid data hazards and other timing-dependent inconsistencies. There is much research left to be done on integrating asynchronous systems with adaptive design, and this will likely become an important area in future nanoscale technologies.

Adaptive and variation-aware systems will be useful in any new technology, so it is important to recognize that these systems are not just for extending the life of CMOS, but for pushing the boundaries of device design in general. Work in variation-aware system design will continue to be important for the entire future of computing.

Index

A

Absolute zero, 2, 15
Activity factor, 66, 108, 132, 134
Adaptive body bias(ing) (ABB), 11, 63, 77,
    78, 96–98, 104, 114, 120, 121
Adaptive voltage scaling (AVS), 10, 11, 63,
    64, 76–78
Al, 24, 110
Ambient temperature, 7, 26, 35, 83
Archimedes, 3
Area overhead, 35, 42, 64, 89, 109,
    113–114, 136–137, 160, 165, 166
At-speed testing, 149

B

Berkeley short-channel IGFET model (BSIM),
    18, 20, 66, 77
Black's equation, 24
Boltzmann constant, 17
Bose–Chaudhuri–Hocquenghem (BCH) codes,
    144, 158–160, 162, 164, 165
Buffer insertion, 117
Bulk-charge Coulombic scattering, 18
Burst error, 145, 158, 164

C

Carrier density, 15–18, 21
Carrier diffusion, 15, 20
Celsius, Anders, 2
Clock frequency, 31, 66, 159
Clock gating, 10, 63, 160
Clock skew, 11, 93, 94, 109–111
Code rate, 144
Comparator, 48, 49, 51, 86
Component temperature sensors, 48, 50–51

Conductivity, 4, 5, 16, 43, 83
Critical charge (Q_{crit}), 65, 142
Critical path replica, 64
Cu, 24, 110
Current density, 15, 21, 24
Current–temperature (I–T) dependences, 47
Current–temperature (I–T) slopes, 26

D

DC–DC converters, 48, 120, 169
Decoupling capacitors, 45, 51, 68, 118, 169
Delay, 7, 25, 35, 63, 64, 66–68, 71–88, 93,
    125, 149, 170
Delay failure, 4, 85, 141
Delay-tracking sensor, 47, 54
Delay uncertainty, 93, 94, 103, 104, 108,
    113, 115, 120
Depletion region charge density at surface
    inversion, 97
Dopant, 17, 18, 65, 169
Dopant concentration, 16, 19, 20, 23, 97
Drebbel, Cornelius, 1
Drift velocity, 18
Dynamic I/O bitline repair, 145
Dynamic voltage and frequency scaling
    (DVFS) system, 81, 83–85, 119, 120
Dynamic voltage scaling (DVS) system,
    78, 104

E

Einstein relationship, 20
Electric field, 6, 18, 20, 21
Electromigration, 7, 8, 15, 24–25, 143, 149
Enable pulse width, 37, 39–41, 44, 45, 50,
    52, 87

D. Wolpert and P. Ampadu, *Managing Temperature Effects in Nanoscale
Adaptive Systems*, DOI 10.1007/978-1-4614-0748-5,
© Springer Science+Business Media, LLC 2012

Enable signal, 36, 38, 40, 87, 129, 130
Energy band gap, 15–16, 18
Energy dissipation, 40–44, 51, 59, 77
Error control coding (ECC), 137–138, 141,
    143–145, 153, 165
EUV lithography, 65

F
Fabrication limitations, 65
Fahrenheit, Daniel Gabriel, 1
Faraday, Michael, 4
Fermi energy, 16, 17, 22
Flatband voltage, 22
Flip-flop, 36, 37, 50, 87, 106, 108, 130,
    132–134, 137, 159
Fludd, Robert, 1
Forward body bias (FBB), 77

G
GaAs, 16
Galileo, 1
Gate-channel work function, 97
Gate leakage current, 24
Gate length, 70
Gate overdrive, 29, 70, 98
Gate-substrate contact potential, 22
Ge, 16
Gilbert, William, 3
Globally-asynchronous locally-synchronous
    (GALS) design, 66, 170
Global temperature variation, 9, 10
Guardbands, 31, 36, 47, 53, 63, 64, 66, 67,
    70, 78–80, 83–86, 93, 142, 170

H
Hamming code, 144–146, 158, 164
Hamming distance, 143
High-κ dielectrics, 10, 15, 28, 29
High threshold voltage (HVT), 95
High-voltage boost for low-voltage
    transition, 126
Hotspot, 7, 9, 137
H-tree clock topology, 110

I
In-line test (ILT) error detection system,
    147–153, 158–162, 164
Interconnect resistance, 15, 24, 68, 109, 110
Interface-charge Coulombic scattering, 18

Interleaving, 96, 145, 158, 164
Intermittent faults, 141, 147
International Technology Roadmap for
    Semiconductors (ITRS), 4
Intra-die temperature variation, 7, 11, 35
Intra-die variation, 65, 89
Intrinsic carrier concentration, 17, 22, 97
Inverting buffer, 117, 133
IR drop, 31, 63, 86, 110

J
Junction temperature, 3

K
Kill-zero carry lookahead adder, 48, 49

L
Leakage current, 8, 15, 22–24, 31, 59, 66
Leakage power, 63, 66, 89, 104
Level converters, 54, 110, 126, 137
Level converters with frequency doublers
    (LCFD), 110, 111
Level shifters, 49, 51, 128, 132, 133, 136–139
Linearity, 56, 57, 113
Locally-asynchronous globally-synchronous
    (LAGS) design, 66, 170
Local temperature variations, 7–8, 169, 170
Look-up table (LUT), 30, 44, 47, 64,
    80–82, 84, 85, 127, 129, 130, 137,
    147, 148
Low threshold voltage (LVT), 95–97, 104
Low-voltage differential signaling
    (LVDS), 126

M
Majority voting, 143
Matthiessen's rule, 18
Mean time to failure (MTTF), 24, 25
Mirror adder, 106, 107, 109
Mobility, 15, 18–21, 25, 27, 47, 93
Monte Carlo analysis, 74
MOSFET, 15, 18, 22, 24, 25, 97
Multi-bit upset (MBU), 142
Multi-V_T design, 94, 96, 97

N
Negative current temperature (I–T) slope, 26
Network-on-Chip (NoC), 125, 145, 165

Ni-Fe thermistor, 54
Nominal supply voltage, 28, 104, 121
Normal temperature dependence, 3, 28, 47, 96, 113, 125

O
On-chip interconnect, 59, 137, 141–166
Open wires, 149
Operating voltage, 28–31, 47, 48, 66, 77–78
Oscillator frequency, 37, 38, 45, 50, 56, 79, 110, 129
Overheating, 10, 47, 59

P
Peltier, Jean Charles Athanase, 3
Percentage-based design, 71–72
Permanent faults, 11, 141, 147, 164
Permittivity, 18, 22, 97
Phase-locked loop (PLLs), 79, 85
Phonon scattering, 18, 20
P/N (β) ratio, 100, 101
Positive current–temperature (I–T) slope, 26
Post-fabrication tuning, 57
Power, 3, 25, 35, 63–66, 70, 73, 74, 76–80, 82–89, 93, 126, 144, 170
Power budget, 9, 64, 170
Power density, 4, 6, 9, 10, 44
Power dissipation, 4–7, 9, 36, 40–42, 63, 66, 70, 83, 85, 125, 132, 135
Power distribution, 79
Power-gated, 43, 51
Power gating, 10, 63, 89, 104, 164
Process compensation unit, 35, 45–46, 57
Process corners, 69, 70, 80–82, 84, 88, 118, 119
Process variation, 31, 45, 51, 57, 58, 60, 63, 65–66, 69, 70, 77, 78, 80–82, 118–121, 169, 170
Programmable temperature compensation devices (PTCDs), 99–101, 104, 109, 113, 115–117, 121
Pulse counter, 36, 50, 87, 129
Pulsed-bus signaling, 126
Pulse generator, 36, 37, 50, 86, 87
PVT variations, 63, 118, 126, 170

R
Reference voltage, 110
Reliability, 4–6, 8, 9, 11, 14, 24, 25, 28, 35, 42, 47, 64–70, 85, 93–95, 120, 125, 134, 137, 141, 164, 165

Reverse temperature dependence, 10, 15, 25–28, 30, 31, 47, 71, 76, 96, 102, 113, 125, 127, 137, 138
Ring oscillator, 36, 38, 41, 50, 51, 58, 64, 69, 77, 79, 80, 82, 85–88, 100, 103, 110, 112–114, 120, 129
Rømer, Ole, 1
Runtime calibration, 35
Runtime variations, 10, 63, 64, 66–67, 69, 70, 76–78, 80, 82, 85, 141, 170

S
Sanctorius, Santorio, 1
Seebeck, Thomas Johann, 3
Safety mode, 10, 85–89, 95, 116
Saturation velocity, 15, 20–21, 25, 26, 93
Sensor latency, 89
Sensor polling rate, 44, 128
Sensor resolution, 35, 36, 38, 44, 50, 51, 79, 86, 129
Sensor sampling rate, 50, 130
Shockley diode model, 22
Short circuit power, 39, 66
Shorted wires, 148–152
Silicon (Si), 5, 9, 16, 18–20, 22, 27, 44, 83
Silicon carbide (SiC), 9
Silicon-on-insulator (SoI), 5, 10, 169
Single-event upset (SEU), 142
Sir Humphry Davy, 3
Slew rate, 109, 115–116, 138, 139
Source-body voltage (V_{SB}), 96, 97
Spare wire reconfiguration procedure, 141
Standard body bias (SBB), 77
Standard threshold voltage (SVT) devices, 93, 95–98, 104, 114, 131
Subthreshold leakage current, 8, 22–24
Supply voltage scaling, 78
Surface roughness scattering, 18
Switching activity factor, 66, 132
Synchronization, 38, 39, 155
Syndrome-storing based detection (SSD) method, 145, 158

T
Technology scaling factor, 4
Temperature-aware coding, 126–127
Temperature-aware delay borrowing, 125, 128, 131, 136
Temperature-aware floorplanning, 9, 110, 126–127
Temperature-aware routing, 126–127

Temperature-aware scheduling, 127
Temperature-aware voltage/frequency
   throttling, 127
Temperature dependency, 3, 15–31, 35–59,
   67, 93–121, 125–139
   of combinational logic, 107
   of interconnect, 109, 116, 117
   of level shifters, 138–139
   of sequential logic, 109
   sensor, 31, 35, 47–59, 120
Temperature-induced timing failure, 54, 95
Temperature-insensitive clock tree, 109–110
Temperature-insensitive voltage (V_{INS}), 25,
   26, 28–30, 53, 67, 93–98, 100, 101,
   103–111, 118, 125, 127, 134, 136
Temperature inversion, 26
Temperature reference sensor, 54
Temperature resilience, 93, 94, 108,
   111–113, 116
Temperature sensitivity, 56, 94–99,
   101–103, 108–113, 115–120
Temperature sensor, 2, 7, 35–48, 50–51, 53,
   54, 57, 59, 64, 78–80, 88, 95, 110,
   112, 119, 128–131, 137, 170
Temperature threshold, 36, 43, 45, 67
Temperature variations, 7–11, 15, 28, 35,
   36, 63–71, 74–76, 78, 80–82, 93, 106,
   110–112, 116, 125–127, 136, 169, 170
Test pattern generator (TPG), 146, 147
Thermal conductivity, 4, 5, 16, 43, 83
Thermal diffusivity, 43
Thermal emergencies, 119, 120
Thermal failure, 4
Thermal resistance, 83
Thermal runaway, 8, 31, 83, 120, 141

Thermal throttle, 95
Thermal voltage, 20, 22
Thermometer, 1, 2
Thermoscope, 1, 2
Thomson, William (Lord Kelvin), 2
3D integration, 169
Threshold Voltage (V_T), 15, 18, 21–23,
   25–28, 47, 66, 67, 70, 87, 93, 95, 97,
   103, 104, 108, 127, 150
Timing failures, 7, 9, 24, 47, 54, 59, 80, 95
Transient faults, 153
Transmission gate, 36, 38, 39, 50, 51, 106,
   107, 109
Triple modular redundancy, 143–144, 153, 160
Tri-state buffers, 160, 165

U
Unit charge, 97

V
Varshni equation, 15, 16
Velocity saturation, 15, 20–21, 25, 26, 93
Very large scale integration (VLSI), 8–11,
   35, 65
Voltage islands, 50, 83
Voltage swing, 66, 125, 127
Voltage variation, 29, 45, 48, 53, 57, 63,
   68–70, 76, 80–82, 86–88, 111, 112,
   117, 118, 121, 169

Y
Yield-based design, 72–73